MILL
POOL

WEED RACK

STREAM

DRIVE

W

N

E

# KINGFISHER MILL

AYLMER TRYON

# KINGFISHER MILL

With a Foreword by
Lord Home

Illustrations by Rodger McPhail

COLLINS
8 Grafton Street, London W1
1985

William Collins Sons and Co Ltd
London · Glasgow · Sydney · Auckland
Toronto · Johannesburg

BRITISH LIBRARY CATALOGUING IN PUBLICATION DATA
Tryon, Aylmer
Kingfisher mill.
1. Natural history – England – Wiltshire
I. Title
574.9423'1    QH138.W5

First published 1985

ISBN 0 00217528 2

Photoset in Linotron Plantin by
Rowland Phototypesetting Ltd
Bury St Edmunds, Suffolk
Printed in Great Britain by
W. S. Cowell Ltd, Ipswich, Suffolk

# Contents

# Foreword

This is the story of a mill house, but it is no ordinary mill house, and for three reasons.

The first is that it has a 'picture' window which looks downstream as the 'race' fans out into a tranquil pool.

The second is that behind the window the author of this book has sat with his binoculars through all the seasons, sensitive to all the sights and signals of nature. He is the most accurate observer.

The third is that the permanent guest is a kingfisher who has clearly understood that his rent for his perch above the stream is to delight and to entertain his host and the visitors to the mill. He holds the stage, not only because he is the jewel among the birds, but because he is the master fisherman, diving and catching his fish and eating them with a speed which almost defeats the human eye.

Aylmer Tryon can sit still (the secret of all observation of nature) and all the birds and beasts and butterflies will come to keep him company.

The heron – named 'Sir Edward' after Grey, the distinguished Foreign Secretary and naturalist; the courting swans, with their necks intricately entwined and stretching to the skies in a love-dance; the grey wagtails, dipping their beaks in the water off 'Daffodil Promontory', then lifting themselves as light and translucent as the thistledown that blows with them on the breeze; and, of course, the decorative mallard and teal.

The reader will be able to make acquaintance with these and many more species of birds, animals and insects, for the shrubs and flowers of the water garden and everything else around the mill cater for their comfort and survival. Clumps of bamboo, for instance, which harbour a myriad of small birds seeking refuge from the winds. Buddleia 'Royal Red', which has been selected because it is the only red variety of that shrub which attracts the Red Admiral and the Painted Lady, the tortoiseshell and the Comma butterflies, among others.

Of course there were hazards in the creation of this paradise. The author, bending down to clean the sluices, was swept by the 'race' through the tunnel into his picture pool; installing bees he ended with one up each nostril.

Aylmer Tryon lived and learned, and he is generous with tips for other pioneers. How to train a pig to find truffles; how to eat fungi without tears; how to talk to flowers, following the example of the lady who told his laggard hyacinths to hurry up; and how, although the result was not entirely successful, to divert a young hedgehog from its apparent intent to commit suicide.

All this and much more is told in this enchanting account of the ways of nature as seen from the picture window in Kingfisher Mill.

LORD HOME

# Introduction

This is the story of a Wiltshire mill astride the River Avon; of my attempts to create a garden from the former water meadow; and of the river life in the watery world as seen from my picture window. The site has much in common with Kenneth Grahame's delightful *Wind in the Willows*; Ratty and Mole are always present whilst the native willows, and the weeping willows which I planted, sway in the wind.

Most bird hides are temporary, often erected for a short time to view particular birds. My mill house is a permanent hide among the resident birds and animals that are my constant neighbours, whilst the migrants are eagerly awaited in both spring and autumn and sadly missed when they have gone. I have been able to study their daily habits and discover for myself much that was previously unknown to me, as I watch the many fascinating creatures unfold their intermingled lives in and beside the river below.

Water, especially running water, has always held a special fascination for me. The reflections are ever changing in all seasons. The sound of the river in flood or the soft summer murmurings of water flowing through stones and weeds is music to my ears, together with the wind in the willows and the call of birds which, sadly, I now have difficulty in hearing.

Water rails and dabchicks compete for the prickly fish known as 'miller's thumbs' with the kingfishers diving from the perch I erected for them. A visiting osprey stayed for about five weeks one autumn and a salmon lived for a short while in the house. There is never a dull moment for those lucky enough to live in such surroundings.

My own interest in natural history was perhaps inevitable, having been born and brought up in the beautiful valley of the Upper Avon.

My other great interest was in pictures, especially of the countryside, which led to my founding a gallery which specializes in sporting and natural history subjects. This enabled me to travel to many parts of the world in search of artists, accompanied, I must confess, by a fishing rod where appropriate.

A friend gave me a small pair of binoculars which I take everywhere, and no present has given me greater pleasure. But most of my bird and animal watching is done from my picture window which is both the lens and the inspiration of this book about my 'elysian' home, ('elysian' being defined in my dictionary as 'a place of ideal happiness').

Before compiling this book I considered the possibility of arranging the chapters by seasons but decided that the characters involved, some of whom are present only at certain times of the year, required individual descriptions. So the seasons and the weather are reflected in sections devoted to them. Nevertheless, living surrounded by water and nature I am always acutely conscious of the time of year and of the changing weather, as the wind drives the rain against the window or the sun's warmth penetrates the room.

The eagerly awaited spring has been heralded by some plants – winter-flowering crocuses, *Viburnum bodnantense*, some daphnes and hellebores – and not too long after Christmas the aconites and snowdrops and other early spring bulbs will begin to force their way through to daylight, and perhaps to snow.

Suddenly spring is here with the first migrants and:

> *This is the weather the cuckoo likes,*
> *and so do I*
> *When showers betumble the chestnut spikes*
> *and nestlings fly–*

The air is full of birdsong and the trees don their spring plumage with the birds; the trout, thin and hungry after their spawning exertions, begin to rise to early-hatching fly. Soon summer is with us with the scent of roses and new mown grass; summer merges imperceptibly with autumn as the trees turn to gold before leaf-fall. And so we must prepare once again for winter, for migrant redwings and fieldfares have replaced the swallows and martins of the spring.

Thus the year rushes by, perhaps silenced for a time by a blanket of snow, which makes my causeway perilous. When the thaw comes:

> *This is the weather the shepherd shuns*
> *and so do I*
> *When beeches drip in browns and duns*
> *and thresh and ply*
> *And meadow rivulets overflow*
> *and drops on gate bars hang in a row*
> *And rooks and families homeward go*
> *and so do I.*
>
> (Thomas Hardy)

# The Village

The diarist Evelyn records that on 22 July 1654, after visiting Salisbury and Wilton, 'We departed and dined at a ferme on my uncle Hungerford's called Darneford Magna, situate in a valley under the Plaine, most sweetly watered, abounding in troutes catched by speare in the night when they came attracted by a light set in ye sterne of a boate.'

Fishing methods have changed since those days and the Piscatorial Society does not encourage its members to try such exciting ploys, but happily the beauty of our village remains despite a few modern buildings. We have our very fine Norman church presiding over the village and many early thatched cottages which for years have been cared for by our local thatcher, Mr Pearce. With the pride and skill of the true craftsman he embellishes the ridges and adds a pair of doves here or a cock pheasant there, so life-like that strangers leaving the Black Horse on a moonlit night have mistaken them for a possible source of dinner. Good thatching straw, plentiful in the days of the sickle and the reaper and binder, is now only obtainable from a few fields reaped especially for the purpose.

We have a well-cared-for cricket ground and some lusty players, especially our wicket-keeper Mick Hazzard who, if he survives the first ball, will often hit the next over the road into his father-in-law's (and our former captain's) farmyard.

I was born in the village of Great Durnford and have lived here whenever possible ever since; I am most fortunate to have grown up in such surroundings, and with such delightful and varied neighbours.

All Wiltshiremen – or Moonrakers as we are sometimes called – have a lasting affection for the downs. For those who may not know, the term Moonraker was supposedly derived from an occasion on a night lit by a full moon – always associated with temporary madness – when a party of men were recovering casks of brandy from a dew pond with long-handled rakes. Excise men suddenly appeared and asked what they were doing to which they replied, 'We be trying to recover that old cheese', as they raked frantically at the moon's reflection. This shows that 'we bain't as daft as us seems'.

> *The Moon does with delight*
> *look round her when the heavens are bare;*
> *Waters on a starry night*
> *are beautiful and fair.*
> (William Wordsworth)

The moon and its various phases throughout the year have always been of special significance to countrymen and in particular to farmers, associated as they are with much lore relating to the sowing of crops and other farming operations, and culminating in the huge yellow moon at harvest time.

At the mill, the full moon shining directly up the river traces a path of special beauty, sometimes known as 'moonglade'. This does not spread as sunlight does on water, but seems to lay a path along the stream directly to the house. If I wake on such a night I go quietly to the window and gaze enthralled by the silvery beauty of this moonlight scene; all nature too seems stilled in wonderment.

The willows, alders and rushy banks etched against the bright light cast deep shadows, and the moonlight shimmers where shallows break the surface and sparkle like diamonds set in a surrounding mirror. All is quiet and even the night birds scarce dare to call. Sometimes clouds are driven by the wind across the surface of the moon as the willow branches sway, but a still quiet night of full moon is to me the greatest enchantment.

I creep back to bed and wake to noisy dawn as the countryside too stirs from sleep and stronger colours are painted once more on the landscape by the rays of the rising sun. My bedroom has all but an eastern aspect so that Clough's beautiful lines, immortalized by Winston Churchill in the last war, seem particularly appropriate:

*And not by Eastern windows only,*
*When daylight comes, comes in the light,*
*In front the sun climbs slow, how slowly,*
*But Westward look, the land is bright.*

When the noisy aircraft and clamorous rooks have returned to their respective roosts the downs have a quite compelling feeling of antiquity. It is not solely derived from the ancient barrows, forts, droves and monuments of mysterious origin, but from the very atmosphere of the ancient turf, of the valleys carved by the ice and seas, now reflected in shadow as the sinking sun reveals their sculptural features where sheep now safely graze.

The downs still retain a diminishing variety of fauna and flora, especially where the steep slopes prevent the ruthless destruction of sprays and plough. Here the ancient junipers still grow; the chalk hill blue and the common blue butterfly, partly dependent on the rest-harrows, now scarce when at one time so plentiful, as its name implies. A great variety of orchids survive and are cherished by some devoted farmers, for fortunately there are many conservationists among them. A few stone curlew and hobbies return with wheatears in spring. Anyone interested in our downlands should read again W. H. Hudson's *A Shepherd's Life* in which his great feeling for this very special environment is so wonderfully portrayed.

From the top of our Jubilee Hill which lies to the east of the village, where bonfires are still lit on such auspicious occasions, the graceful spire of Salisbury Cathedral may be seen. Though built in the low-lying water meadows, this emblem of Wessex dominates the surrounding countryside and the hearts of those who are fortunate enough to live within its aura.

The spire, which was added some hundred years after the early thirteenth-century cathedral was itself built, has to be restored at enormous cost. It is hoped that the appeal will reach far beyond those who live in Wessex since apart from the cathedral's religious significance it is also a monument to the incredible skills of the architects, masons and other craftsmen who added this glorious edifice to the existing building over six centuries ago.

The Close with its many beautiful houses is a fine setting to the jewel itself. I have twice recently seen the image of spire and tower projected by the floodlights below onto low clouds as if on a screen – a strangely awe-inspiring spectacle.

We are fortunate that John Constable, England's greatest landscape painter, found inspiration in his views of the cathedral from the meadows, and we trust these will remain for ever as he saw them.

# History of the Old Mill

Whilst the main purpose of this book is concerned with the bird, animal, and fish life seen from my permanent 'hide', it is interesting to reflect on the origins of the mill itself and the water meadows which surround it.

I am indebted to Miss Margaret Briggs of Woodford for the information on the water meadows and to Mrs Perret of Wilton for her research on the mill itself in the course of her study of the water mills in this part of the Avon valley.

A mill is mentioned here in the Domesday book and no doubt the mill leat was dug on dry land and the river channel diverted into it. Mrs Perret surmises that the present main structure was built in the early or mid eighteenth century and that it was added to and altered as required by the millers, according to what they could afford – a mill is shown here on a map of 1773. She aptly describes the structure and mill furniture as hybrid, the material used being local flint, and chalk clunch known as 'poorstone' set in squares in the attractive Wiltshire fashion of that time. The gearing wheels and the sails of the water wheel and cogs were of wood, gradually replaced by cast wheels as they wore out. The axel tree is probably earlier and may have come from a ship's mast or timber. The roof tiles of baked clay were cheaper than slates and were restored to the roof in my present building. The 'floor' beneath the water wheel was constructed of chalk, as I discovered when wading beneath the hatch, when my foot disappeared into a hole worn by a hanging chain. The chalk was protected from frost by the water.

The original grinding stones have been discarded, but I remember them being of compressed chalk with lead lumps for additional weight, and stone-faced. Mrs Perret deduces from my inadequate description that one pair would have been of French

Bhur for grinding wheat and the other of Peak or Derby for oats and barley. I can just remember them in use, and the clatter of wheels that shook the whole building.

The date of the water meadow system around the mill is probably mid-seventeenth century. Some believe that these meadows were laid out by Dutch engineers who, it is thought, planned the system along the whole course of the River Avon. Such engineers had earlier drained the fens against much opposition from the local people.

The mill has always held a singular appeal for me. As a small boy I would frequently visit it to watch the grinding of corn; climb up into the lofts where some of the corn was stored; chase the mice which abounded, and peer into the pool below. This was usually inhabited by large trout which, as I grew older, I would try to catch as they dropped downstream to chase some passing fly or minnow.

As the horses and wagons were replaced by tractors, so the reaper and binder gave way to the combine harvester and with them the sheaves and stooks where we had assisted – if such it could be called – by chasing the rabbits as they bolted from the ever shrinking corn. A cornfield with the stooks casting long shadows as the sun sank over the downs is a lasting and beautiful memory, and a sight long lost to the eye, whilst the melodious tinkling of the sheep's bells as the flocks moved along the grassy downs is a tune long since unheard. Mechanization has destroyed so much that was beautiful at harvest time, and as the grinding of corn by water power vanished too, the sound of the turning mill wheels and grinding stones mingling with the roaring waters became only a haunting medley of memories from the past.

The mill stood sadly, empty and unused and, as with all empty buildings, soon began to decay. My brother, who owned the mill, had the old tiles removed and stored but soon the chalk and flint walls began to crumble. Fearing that small boys might find fascination in playing there, and be injured, he decided to pull down the building. At this point my sister-in-law suggested that the excellent New Zealand architect Ralph Blenham Bull, who lived in the Close in Salisbury – and what more inspiring place for an architect to live? – should be asked to draw up plans for its conversion into a possible dwelling. I shall always be indebted to her for this inspired thought.

I was later shown the plans and at once agreed to take on the project. A bridle path and right of way passes close on the north side but the architect's plans solved this problem and that of possible flooding – an occupational hazard – by elevating the house on stilts as it were so that the front door was on the first floor level. This led straight into the sitting-room, dining-room and kitchen, with bedrooms above. The ground floor at river level housed the garage, the rod room and the larder and was thus designed to keep the rest tidy – something I am still trying to achieve.

The architect and I had few disagreements; we avoided them by adopting a sort of barter system. Thus, I insisted that the large picture window should look straight down the river, whilst I gave way on some point such as a glass panel in the sitting-room floor through which I would have watched the trout below. He pointed out that I could observe the river from all the windows in the room; moreover that glass conducts sound and he had gone to considerable trouble to lag the floor not only for warmth but to suppress noise, so that only the soporific sound of water could be heard.

The picture window has proved to be in every sense the highlight of the house. When alone I have breakfast there, as often as not watching the kingfisher eating his fish. My guests are invariably drawn towards it as if by some magical spell and even after nightfall I can switch on a floodlight to illuminate again the shimmering water and swaying branches of the willows. My excellent architect may have erred in his dislike of such a large glass area which, seen from below, threatened to spoil his beautifully planned facade, but it was his inspiration which raised the room so high above the river and thus provided for our delighted gaze the view of the incessant river life.

The architect retained the original height and shape of the old mill. The weatherboarding on the southern aspect has indeed weathered; the climbers which I planted delighted in the moisture and soon covered the house to such an extent that the wisteria

now poses a threat to the tiles. The willows which I put in nearby and the surrounding alders now help to give the mill an appearance, if not of antiquity, then at least of maturity.

The conversion of the house was entrusted to Mr Waite and his two sons; to them, and especially to the architect, I shall always be grateful. The house grew rapidly and the old roof tiles were finally restored to their original position. The old walls of flint, chalk and brick had to be reduced to a height of about four feet so that a damp course could be put in. The work was completed in the spring of 1961.

I am most fortunate in having a painting of the old mill as it was just after the war. It happened in this way. A great friend of mine was killed in the war and left me some money. I was an admirer of the work of Edward Seago and so I wrote to him to ask whether he would be kind enough to come down and paint a picture for me. The summer was very hot, the river low and the downland grass parched – almost the colour of grass in the Australian outback. The artist, whom I had not previously met, regaled us with amusing stories and comments on the difficult art of land-scape painting and choice of scene. He liked to sketch small panels in oils which later he sent to me so that I could choose the final picture. I chose one of the village painted from the top of our Jubilee Hill with our fine Norman church and many thatched cottages. Beyond the village lies a wooded hill and above it great thunderheads tower to relieve the thirsty fields below. The final result was all that I could wish as a memorial to my friend.

Now my brother was a keen racing man and each year we used to try the Tote Autumn Double. I left the choice mainly to him, but added a couple of horses whose names had some connection with rivers or watery things like fish. It so happened that one of the selections – almost certainly his – had won the first leg so when I wrote to Ted Seago, a friend by now, I said that I would like a 28 × 36″ canvas of the village, and that if one of the following five horses won the second leg then please could he paint me a 20 × 24″ picture of the mill. One of our selections did win and the picture of the mill from beyond the mill pool now hangs above my mantlepiece. This happy event occurred long before I thought of living here.

As the object of this book is primarily to describe the life surrounding the mill I will not linger too long over the contents of the house, but I must mention my mural. I bought a panel painted by Chinese students for the staircase walls. Mandarin ducks, willows and some strange exotic birds mingle with a pair of kingfishers, one with beak pointing upwards and glaring with a furious expression at her mate descending at great speed; this one I call 'Sorry I'm late'.

The various artists who paint for my gallery kindly come to stay from time to time and the happy thought occurred to me to ask them to paint on this mural something appropriate to the mill. The first was Paul Jones, the distinguished Australian flower painter, who enhanced an existing tree peony and on a later visit added a yellow iris or flag, which abound here. Axel Amuchaste-gui, the great bird and animal painter from the Argentine, painted a red squirrel – alas now replaced by the destructive greys. The majority painted direct on to the wall but those who had not the time to do so sent cut-outs: Keith Shackleton for instance contributed an osprey for the ceiling and included instructions for fixing it. The Canadian artist, Robert Bateman, sent three water vole sketches, one with attendant ripples and, on a later visit, painted a charming wren with moss beside the water. Rodger McPhail likes to balance precariously on a pile of logs. He saw a water rail for the first time below the house, did a quick sketch and then, without further reference, painted the bird direct onto the wall. The result was perfect. (As I write one has walked out from the bank and is fishing among the stones, probing for 'miller's thumbs' or water shrimps.) Susan Crawford could hardly be expected to paint a horse so has added a green elf sitting on a willow fishing for Peter Scott's dabchick which swims below. Jack Harrison has painted the essential kingfisher and Margaret Mee sent from Brazil a charming marsh orchid painted from a photograph of one in my meadow. The head of an otter by Ralph Thompson peers hungrily at a nearby trout. Many more are promised and I am most grateful to them all for their skill and the pleasure which they give.

# The Garden

Whilst the house was being built and after the plans had been finally agreed with the architect, my thoughts turned to the possibilities and problems of creating a garden. I read gardening books, consulted friends of mine who owned beautiful gardens, and joined the admirable Royal Horticultural Society, visiting their fortnightly shows and many garden nurseries.

George Taylor, then Director of Kew Gardens, came down to stay and said at once that the surrounding water meadow, withy bed and wilderness was the last sort of place to make a garden, since nature had done so already! Jim Saunders, then head gardener of the Leckford water garden beside the Test which George Taylor had originally helped to design for his friend John Spedan Lewis, came to give advice and later I obtained many plants from this most beautiful water garden, perhaps the most beautiful in England.

I was thus very well advised, but with my complete lack of knowledge I fell eagerly into all the beginner's traps; I ordered plants which appealed to me, forgetting about soil, aspect, and the very wet conditions and, just as with bee keeping, I learnt the hard way.

However, I did have some lucky breaks. Perhaps the most important of all was my friendship with Lionel Richardson, 'king' of the daffodil growers, whom I used to meet fishing in February on the beautiful Blackwater River. We fished there together for many years; sadly he died while I was building my house. His widow sent me a huge sack filled with a great variety of bulbs which he had cultivated over the years and some of which had won many awards. There could not have been a better memorial to him: they spread and multiplied each year, delighting in the moist soil, and every spring the wonderful sight of them in their watery setting serves as a reminder of Lionel and our fishing. Daffodils are indeed courageous plants, as Shakespeare appreciated: they 'come before the swallow dares and take the winds of March with beauty'. I have so often seen them collapsed under snow or frost in early spring but as the sun rises, so too will their lovely flowers. At times of spring flood from melting snows or heavy rain, one daffodil clump at the end of the peninsula is sometimes covered so that only the flowers remain above the surface, waving with the current, not the frantic wave of a drowning man but the cheerful sort, beckoning 'Come on in, the water's fine'. This morning at the end of April, this same clump appeared to be suffering from convulsions: a pair of moorhens were building their nest in the centre! Later they hatched six young, but sadly the parents seemed to have quarrelled, as many do; one, presumably the cock, looked after two babies and the other, four.

Whilst on the subject of daffodils, we have a few at our fishing hut on the River Frome, and each year at Easter we hold a competition with our delightful 'Opposition', who fish the other bank. There are three classes: the largest number of varieties; the finest individual bloom; and, finally and the most hotly contested, the scruffiest daffodil. The first year the prize consisted of a rosette most kindly presented by the opposition's then small daughter Vicky, one of her most valued possessions, bearing the legend 'Aldershot Show, Highly Commended'. This now much faded trophy hangs in the hut of the Victor Ludorum. The rosette for the scruffiest bloom is a golden harp emblem for some equally obscure reason and was presented by John Ashley-Cooper, a man more associated with angling than horticulture.

The first class is always, after much contention, a draw at four varieties, although some ten hopefuls are exhibited and six eliminated. The finest individual bloom is carefully selected and judged. Class Three was the most exciting ever on one occasion – the competition that year was to be held at the opposition's hut on the Rīve Gauche. We had selected a bloom which had been slightly damaged by a scythe and then singed whilst clearing rubbish and as reserve, just in case there was an objection and steward's enquiry, we took another splendidly scruffy bloom along. However, when the Opposition turned their punt the right way up for the first time since its winter repose, there was the smallest, palest and most miserable daffodil we had ever seen, one which had never seen the light of day. We had to concede their victory with the proviso that *Sir Bodkin* – for such was the boat's name – should never again, even by mistake, cover ground which might harbour a bulb.

I have planted many crocus bulbs but mice seem to follow behind, so the next morning only empty holes remain. I even tried putting holly leaves above each bulb as a friend suggested! The only remedy with which I have had some success is to sprinkle Jeyes Fluid about so that the mice are unable to smell the bulbs.

A friend who lives in Portugal gave me some bulbs of *Sternbergia lutea*. These have never flowered, and when I consulted the RHS book on bulbs I learned that this variety (*fisheriana*) seldom if ever flowers here: it only flowers in the wild in Mohammedan cemeteries in Kashmir!

26

I thought that a small herbaceous border would be attractive and would also provide flowers for the house. One day I was walking up the village street when I noticed that a drag-line excavator was widening the road close to one of the best gardens in the village. Someone had once told me that roadside verges were excellent soil and this, with the proof of the garden beyond, seemed just what I sought. The high-sided truck into which the soil had been loaded was just about to move off, so I asked the driver if he could dump it beside my new house. He may have looked rather surprised but he followed me down and I pointed out where he should tip it. He reversed into position and the lorry began to tilt. Just as I was imagining the rich avalanche of loam there was an ominous rumble and out fell the village street – asphalt, rubble and all. That site is now my car park!

I had to remove the mill wheel and grinding stones and other relics, as they had fallen into disrepair. Without wishing to achieve 'Ye Olde Mille' effect, I decided to erect the mill wheel with the idea that a wisteria could grow up the ancient ship's timber which formed the shaft and could then spread itself out along the spokes of the wheel and cascade over the outer rim. We had great difficulty in erecting the wheel, which was of consider-able weight, and at one moment we all let go leaving only my nautical neighbour to take the entire strain. This he did most gallantly until we noticed that his face was turning from purple to blue – almost the colour of the wisteria's flowers – at which point we rushed to his aid and the wheel returned to a horizontal position.

Unfortunately the wood was full of woodworm holes and the wheel soon fell to pieces leaving only the shaft and a few spokes. I then erected poles around the perimeter joined at the top with cross pieces like an octagonal bird cage and we built a flint and chalk wall round the base in Wiltshire fashion, which we filled with earth and spring bulbs. I planted clematis up each upright pole and we wagered on which would first reach the top – Nelly Moser was a firm favourite but was short-headed by the Comtesse de Bouchard, with *Henryi*, the lovely white clematis, in third place. *Macropetela*, the early lavender blue clematis with nodding bell-like flowers, is my favourite.

About this time I saw the beautiful Clematis *orientalis* at Chelsea and ordered several plants. In due course they arrived and I gave them little ladders up which to climb. These they ignored until eventually my brother's expert gardener happened to be passing. 'Why do these clematis refuse to climb?' I asked him. He took one look and said, 'Because they are Christmas roses.' The ignorance was mine but the error was the nursery-man's since he had sent me *Helleborus orientalis* by mistake. This proved a most happy error since the hellebores loved the wet, and later I planted other varieties which flower from early January until May. Sadly the flowers are short-lived in a vase; like many others they are better left in the garden.

A small fast-running stream, formerly part of the water meadow system, borders the east side of the garden and beyond it I planted a beech hedge as a shelter belt, which grew fast and thick. I never understand why such hedges retain their leaves throughout the winter while beech trees themselves shed their leaves, but the hedges give fine autumn and winter colour until the new and the beautiful soft pale green shoots arrive in early May. Here I tried to create a water garden of primulas, water irises, kingcups and other plants which love moisture, and now I plan to plant water lilies in a stagnant back stream.

My most short-lived purchase from the Leckford water garden consisted of about half a dozen of the white, feathery *Smilacina racemosa*, some hostas, and a weigela, which I planted in a group. The following weekend a friend was gazing from my window when he suddenly shouted 'Look out – a rabbit is eating your new plants.' I rushed down, seized my gun and crept round my bamboo clump beside the stream. The rabbit bolted and I shot it but when I picked it up I found that I had also shot all the *Smilacina racemosa*, the weigela, and that I had peppered the hostas so badly that they never recovered. Rabbit pie was really no compensation.

My garden is so wet, being well below river level, that weeding of the primula beds is most difficult, as liquid mud sticks to whatever weapon one might be using. However those primulas which survive, especially the tough *P. florindae* and *P. japonica* and various hybrids, are most rewarding, and some seed and establish themselves. Hostas of various shades look well beside the stream but are subject to depredation by an unknown assailant known as *Hosta muncha*. There are various possible culprits: my numerous and delightful water voles which feed mostly on water weed which they eat, sitting up, with their little hands are the most likely; the sinister slimy slugs are a probable; rabbits are a possibility, as are the cows which once invaded the garden and destroyed in a few minutes the lower branches of my only swamp cypress. However there were none of the footmarks near the hostas which were all too obvious on the lawn (or in the lawn, since they had sunk in at least a foot) so Dr Watson pronounced them innocent. There are some wild flag irises and clumps of kingcups; I have also tried *Iris kaempferi* but there is too much chalk in the water, although *I. sibirica* and *I. laevigata* add their beauty to the reflections in the stream.

The prevailing sou'westerly wind blows strongly up the valley so I planted a bamboo screen of the *japonica* variety much loved by the birds for shelter and nesting. One evening a large number of pied wagtails settled in the willows above, before going down into the bamboos as dusk fell. That night a gale blew up which would have been most uncomfortable in their normal winter reed-bed roosts. Birds certainly have forewarning of wind and hard weather before our weather experts. In very severe weather wrens will roost in tight groups for warmth. During one cold spell a group of about fifty wrens flew each night into the eaves above my bedroom window through a small hole, later usurped by a breeding pair of great tits.

Grey wagtails delight in shallows where the stream runs swiftly and where there is an abundance of fly life; a pair of them nested every other year for six years in the same nest, precariously perched on a very narrow ledge in the room from which the mill wheel had been removed. This made for a most hazardous maiden

flight for the young. They had to fly down to the foaming water below the hatch and then flatten out to fly through the opening below. On one occasion I opened the door and the young flew into the house. The flight of a family of these delightful birds as they dance above the sparkling waters with tails outstretched, and yellow flashes as they catch the hatching flies, resembles the pirouetting of a corps de ballet.

Magnolias are my favourite shrubs. The first and most useful book which I bought was *A Chalk Garden* by F. C. Stern, so I planted *Magnolia highdownensis*, named after the garden which he created at Highdown in Sussex. This beautiful shrub has large white hanging flowers with deep red-purple centres, and the seed pods are coral coloured. Alas, after growing fast for some ten years and flowering freely, a branch died each year until all that remained was the stump, with a single, slender branch which most gallantly produced a final perfect flower – a sad farewell!

At about the same time my bamboos all died; I learnt that the *japonica* species always dies after flowering, though mine were only about fifteen years old. The canes, whilst certainly useful, are no recompense for the many birds which had sought shelter and nesting sites in the thicket. However a magnificent magnolia, Leonard Messel, which has many large pink stellata-shaped flowers and which had hitherto been protected by the bamboos was now more able to look after itself.

I planted *Magnolia grandiflora* on the north wall but sadly this has been one of my many failures: after ten years it is neither *grandi* nor covered in *flora*, whilst its neighbour *Magnolia denudata* is all too eager to produce its large white goblet-shaped flowers before the risk of frost has gone. A friend told me that I should spray the flowers with anti-freeze, so when I woke on a fine but frosty morning in early spring and saw from my bedroom window that the blossoms were covered with hoar frost, I dashed down in my pyjamas, took the can of anti-freeze from my car, and squirted each flower, being just able to reach the topmost. Another friend later told me that all I really needed to do was to squirt it with water – a rather damping remark!

Albertine is perhaps my favourite rose even if it flowers but once in a season, and ramblers and climbing roses, such as the deep red and sweet-scented Etoile de Hollande, seem to thrive on the drips from my veranda. Shrub roses at first protested at their damp feet, but are at last establishing themselves and no longer complain, whilst the white-flowered Filipe fairly dash up the willows as if to avoid drowning. Most of my roses come from the wonderful Cranborne Garden and are a constant delight in summer. Spring is my favourite time of year and so I plant the early sweet-smelling flowers, Viburnum bodnantense, *Daphne odora* and the like.

Lately I have become more interested in autumn flowers, partly because butterflies have always fascinated me and a garden is surely incomplete without their beauty. So I have planted buddleia, 'the butterfly bush'. I do not care for the colour of the common or garden *B. davidii* so I planted some of the many varieties available of more pleasing hues such as deep crimson and white. I now learn belatedly that these are not so popular with butterflies as the true *B. davidii* and so I will revert to this original and hardier shrub even if the colour offends my eye; it will not do so when covered with tortoiseshells, peacocks, the migrant Red Admirals, Commas and my favourite Painted Ladies. I must add that on a recent visit to Hidcote I bought the 'Royal Red' variety of buddleia and on its most informative label I read 'attractive to butterflies' – and so it is! Sedums, Michaelmas daisies and lavenders attract my bees as well as butterflies.

I have just bought Miriam Rothschild's *The Butterfly Gardener* which is fascinating, and as I read I am compiling a long list of plants which I shall place in Isabella's Orchard, my so-called 'conservation area', a wilderness of nettles, brambles and the like. I fear the despair of my good and kind neighbours with their well-cared for and wonderful gardens, but perhaps I will be forgiven my windblown seeds if some of my butterflies alight and grace them with their beauty.

A neighbour found many years ago that there was a silt deposit formed no doubt by the old course of the river but now for the most part well above flood level. This has a fine beech belt, and a path, perhaps at one time a carriage drive, runs for nearly a mile almost to my mill. The path itself in early spring is strewn on

either side with snowdrops and aconites, a wonderful sight when the sun shines and filters through the beech canopy to flood the carpet of flowers beneath with colour. Here were planted camellias and rhododendrons, which flourished in this lime-free leafy sanctuary. I tried planting such shrubs on my island, but there is more gravel and little silt, so the few azaleas and couple of camellias that I put in were miserable. However, to the east of the stream the old water meadow has little chalk and a good depth of soil with a layer of clay about a foot down. When we dug a hole recently in order to remove clay to re-face the kingfisher nesting bank, we reached water only a few inches down; this meant that many plants which at first liked the lack of lime eventually succumbed to the wet. Magnolias certainly prefer the slightly drier places, and in times of flood the chalk stream forces those plants which prefer acid soils to show their resentment by displaying yellow tinged leaves in silent protest.

## CONSERVATORY

My little conservatory which sits rather precariously on the balcony on the east side of the house has never fulfilled my hopes. I gave it an elaborate propagator with a thermostat but somehow the seeds never became seedlings and the sand which covered the floor became a 'dying desert'.

The slatted shelves were covered with trays of sand on which stood pots of various plants. Rainwater from the roof filled a tank and thence irrigated each tray in turn. I have seen the same irrigation system in farms and forests in Chile, though on a somewhat larger scale, which used the melting snows of the Andes! Mine either dried out or flooded the whole floor. Ventilation was provided by an amazingly cunning device which opened and closed two windows at set temperatures. Unfortunately, after a successful start the machine seemed unable to tell hot from cold and I would be showing a friend a withering plant on a very hot day when there would be an alarming noise from above and a clank as the windows closed. However a hoya and plumbago of beautiful blue have survived, and now a plant which has sat sadly on the shelf for so long that I have forgotten the kind donor has flowered in a beautiful violet bloom which I recognized with surprise as Tibouchina from southern Brazil.

My most precious plants were rare orchids, also from Brazil. I visited Margaret Mee, the distinguished and intrepid flower painter and collector in Rio, and brought back some rare orchids from Amazonas and the Matto Grosso for Kew Gardens, and a few for myself. I had been told that it is preferable to grow orchids on their own; in my conservatory I was inadvertently achieving this all too rapidly. As I feared, the orchids never thrived, with the exception of *Laelia purpurata*, which Margaret Mee had illustrated in her book *Flowers of the Brazilian Forests*. She describes it thus ' . . . generally considered one of the most beautiful and elegant of Brazilian orchids. For this reason it has been collected intensively and few are now to be found in their native habitat.' This orchid is known as Queen of the coastal forests. She adds, 'Unfortunately these wonderful coastal forests are being steadily destroyed, often for use as firewood! With them perish untold numbers of animals and plants. So the advance of civilization and industrialization proceeds at the cost of the total destruction of all that is most beautiful and precious in creation.' Later she wrote to say that she was delighted to hear that the orchid was flowering in Wiltshire as it was now even more of a rarity in Brazil. The story has a happy ending since a friend and neighbour, seeing the unhappiness of my orchids, rescued them and they now flourish under expert care in his beautiful orchid house, where, as I write, *Laelia purpurata* is once again in full flower with many more blooms.

I later added a sun room on the south balcony. Here my housekeeper, the indefatigable Mrs Chitty, raises geraniums with loving care and such success that in summer the balcony and stairs up to the front door with their many pots resemble a Spanish patio and bring cheerful colour to the approach. She attributes her success to talking to the plants. This is at times rather confusing. I was coming down to breakfast on time one morning when I suddenly heard her say, 'Hurry up!' I looked to see the cause and found that she was in fact addressing a bowl of hyacinth bulbs which were supposed to flower by Christmas. This reminds me of another occasion when a man was exercising his two black woolly dogs along the bridle path which passes my door, and noticed that one was missing. 'Where's your bloody brother?' he enquired. At that precise moment a friend who was staying with me emerged and, although surprised at the question from a complete stranger, replied unhesitatingly 'Which one?'

My vegetable patch is presided over by Derek, the industrious gardener who comes once a week with the enthusiastic and equally hardworking young Richard. Together they mow, prune and tend the rest of the garden, all in a most cheerful manner although rather disapproving of my affection for the wilder flowers – willow herb, purple loosestrife, kingcups, flags and even daisies and the like. Naturally I like to grow my favourite vegetables – early potatoes, green peas, broad beans and especially sweet corn – whilst Mrs Chitty hovers over my courgettes and pounces on them like a kestrel on a mouse before they become marrows, which seems to occur almost overnight. She then converts them with her considerable culinary skill into courgette soufflé.

Isabella's Orchard, named after my first queen bee whose unfortunate demise is described later, has not so far been a success. I planted Cox's, my favourite apple, but the few so far produced seem to have a different taste. My sister-in-law informs me, quite rightly, that I should plant another variety for cross-pollination. I also planted two varieties of plums but they have yet to 'plum'. Perhaps the orchard is too wet, perhaps the bullfinches nip off the buds in their tiresome and seemingly aimless manner – or perhaps this year the trees will be laden with fruit worthy of Isabella. Of the larger trees which I planted the yellow-barked willows grew at such an alarming pace that they are now far higher than the house. Seen from the downs above, the boughs have a pleasing pink hue against their background. Their main failing is that in each gale branches fall, often onto unsuspecting shrubs below. My weeping willows, after a poor start when they suffered from a blight, are now fine trees, especially those above the house, and when the pale green shoots first break they seem to trail their fingers in the river like girls from punts at Henley Regatta. A whitebeam with the impressive name of *Sorbus aria majestica* flourished on the island for some years but is now ailing. Whilst fishing on the Spey a few summers ago we noticed in the evening that each of the larches on the distant skyline appeared to be on fire, with a column of smoke wavering in the slight breeze. None of our party had seen such a sight before. Obviously they were insects, perhaps involved in some mating dance like perpendicular mayflies. Last summer I saw a similar column arising from my ailing whitebeam. Those insects were obviously much smaller since I had to walk beneath the tree in order to observe them properly. I fear that it might have been some ritual death dance for a departing tree.

I also planted a swamp cypress which was delighted with its watery surroundings until, as I have already mentioned, opportunist cows rushed through a gate that had been left open and greedily devoured the lower branches. As if this damage was not enough the cows then danced round the poor tree as if performing some ancient tribal sacrifice or more joyful maypole dance, their feet sinking deep into the so-called lawn. Now the tree has almost recovered and has been joined nearby by the kind gift of a soft feathery *Metasequoia* of similar appearance and preference for damp.

## FUNGI

I am very fond of mushrooms and the meadows at one time provided a fine crop each morning. Sprays and ruthless ploughs have destroyed almost all of them, so one morning after an abortive search for mushrooms I picked a large grey fungus with a scaly top, and brought it home to identify. This for once proved easy: the Parasol Mushroom, so named for its shape. It was described as 'edible and excellent' and so it was! This autumn I found a similar-looking mushroom which, after careful comparison with my book, I identified as the Shaggy Parasol. It was equally good to eat. Since then I have acquired several books, one given to me with the inscription:

*'You are much too young to die*
*from eating* Amanita *pie.'*

I was taken on a fungi-gathering expedition to the New Forest and there shown the deadly *Amanitas*, the 'Death cap' of harmless appearance, the glistening, white-topped 'Destroying angel' and the poisonous but not fatal *Amanita*, the 'Fly agaric', with its pinkish scales above easily identified by the gnomes which delight to sit thereon, especially in suburban gardens! 'Lawyers wigs' are easy to recognize and good to eat; once I found some in Green Park in London but these I did not pick! Chanterelle and the edible *Boletus* are excellent and even the young puffballs.

Many people refuse to eat any fungi except the field mushroom but these, after all, are fungi too; I think the golden rule is to learn to identify the dangerous kinds, then the easily recognized edible fungi, and never to experiment on yourself, on friends, or even on enemies!

Last autumn a man walked down the path past my house with his small son, carrying a couple of fungi. I asked to see them and one was almost certainly the 'Yellow stainer', which resembles the horse mushroom but turns yellow if bruised. This fungi is harmless to some but poisonous to others and a neighbour was ill for several days after eating some, so I advised the passer-by to throw them away – for all he knew they could as easily have been the fatal 'Death cap'.

Truffles are a form of fungi which grow under the surface beneath beeches and oaks. My father told me that a man from Winterslow used to come with his truffle hounds to seek for these delicacies, so highly prized and priced in France and Italy for their delicious flavouring.

When a friend returned from Italy where she and her husband had been instructed in the art of truffle-hunting with a pig, they brought a truffle back with them and waved it over a litter of piglets until one showed an interest and tried to grab it. They then bought this piglet. I suggested that perhaps the Durnford truffles had multiplied, so one wet November evening they arrived with the pig and a terrier. I was not clear as to the part that the terrier was to play but the animals certainly did not like each other and the pig screamed at the dog. The local saddler had constructed a harness rather similar to that made for the excellent guide dogs – he had not seemed at all surprised by his unusual order. However,

when the harness was fitted to the pig, which was about one third grown, it gave a display worthy of any bucking bronco and finally dashed off into the now damp and darkening wood dragging his owner through the nettles at high speed, whilst continuing to scream in loud protest.

By this time several locals, attracted by the noise, had gathered; some of us had been armed with forked probes. We followed in a disorderly rabble stumbling as we went, and whenever the pig, which had by then steadied to a trot, checked to sniff the ground, it was restrained whilst we dug hopefully. The hunt was called off when darkness made truffles difficult to spot and the followers were too nettle-stung and damp.

The fact that no truffles were found was variously attributed, by some to an absence of any such things, by others to the fact that few of us knew what we were looking for, or that perhaps the pig had surreptitiously eaten them before being restrained. From the pig's smug expression I rather supported the latter view. Even if our dishes lacked that unmistakable and delicious flavour, the Great Truffle Hunt was considered a great success.

## BEES

*. . . and still more, later flowers for the bees*
*Until they think warm days will never cease*
*For summer hath o'erbrimmed their clammy cells*

(Keats)

I decided that I would like to try bee-keeping simply because, like Pooh, I am very fond of honey. Long ago the son of a friend, then aged about six, was given a most generous tip by an American godfather. 'What will you spend it on?' I asked him. 'A birdcage,' he replied. 'But that is useless without a bird,' I said without thinking. 'Yes,' he said, 'but better than a bird without a cage.' I learnt my lesson then.

I found that a friend of mine had a spare hive, and so I drove down to Ferndown to call on Mr Vicary of South Coast Honey Farms. I found him in a shed surrounded by an amazing collection of bee impedimenta and indeed by a quantity of bees which I eyed with some misgivings. I left with a variety of trappings and with Isabella, my first queen, and her entourage in a rather flimsy wooden box containing about six frames known as a nucleus.

Isabella was a young queen and had been adorned with a green spot which denoted that particular year. I drove home with considerable care since a sudden application of the brakes would surely have put an end to my bee-keeping aspirations if not to me. On arrival, Isabella settled down well. I had been told that there

40

was no fear of her swarming in her first year. However some weeks later Mrs Chitty, telephoned to me in my London office to report that Isabella had swarmed. I therefore telephoned to Allen Cook, an expert bee-keeper (and fishing-keeper) on the Upper Avon. He was out but his wife promised to give him the message on his return. This I reported to Mrs Chitty.

Some time later a van drove along the track leading to my house. Mrs Chitty, who had been hovering anxiously, rushed to the driver exclaiming 'Have you come to rescue Isabella?' 'No,' he replied, 'I have brought some wine for Mr Tryon from Portugal, and who is Isabella?' Having delivered the wine he was driving away when it occurred to him that he might be able to help, so he asked where Isabella might be. On being told that she was up a cherry tree and having realized by then that she was a queen bee, he asked for a box and ladder. Paul Gorriup, for such was his name, had been on a great bustard mission to Portugal. He had never before taken a swarm, but being a naturalist he knew that bees are supposed to be comparatively harmless when swarming and so very bravely shook the swarm into the box and from there poured it back into the top of the hive. About an hour later Allen Cook appeared and said that whilst this was not the correct practice, Paul had saved the swarm which naturally would take some time to sort itself out again. He then went through the hive and removed the two young queens that were the cause of the trouble.

During Isabella's first and, sadly, last winter we had a flood and I received a message from Mrs Chitty, whilst I was having lunch with a neighbour, that Isabella's palace was in danger of floating down to Salisbury. I returned at once to find that the waters had risen to the very foundations – a concrete slab of pinkish hue on which the hive stood. Fortunately at that moment a temporary neighbour, a serving officer and a brave man, came splashing along the causeway. He said he had little experience of bees but offered to help. So we lifted the hive and carried it – most precariously, since the meadow is rough and undulating and we could not see where we were putting our feet. We had one alarming stumble but reached terra firma safely. Bees in mid-winter evidently sleep very soundly. That year provided a short early spell of fine weather. I had just given Isabella's hive a large dish of candy and stupidly thought that this would last them some time. When next I looked all had died. I planted an orchard in her memory and as a penance for my ignorance.

My next queen was Boadicea who was given to me, with her swarm, by Allen Cook. She was usually good-tempered and evidently gave instructions to her guards that they were to be firm but gentle with intruders. However the weather had been inclement and even the nicest bees become restless and irascible if confined to barracks. I stupidly decided to inspect the hive and, arming Mrs Chitty with a bee bonnet and gloves, I lifted off its roof. I was aware at once that the bees were displeased from the roaring crescendo like that of an underground train emerging from its tunnel into a station.

I had armed Mrs Chitty with the smoker which, if it hasn't by then gone out, is puffed into the top of the hive, in theory causing the bees to seek shelter from possible fire in the nether regions of the hive where they hastily gather honey in case the order is given to abandon ship! On this occasion, having been given orders by a

petulant Boedicea to attack, they did so with considerable effect. Several got through Mrs Chitty's defences and on being stung on the ear and head she quite rightly fled but, in so doing, dropped the smoker. When I stooped down to retrieve it some of the guards, who had evidently been awaiting just such a moment, squeezed under my veil. Until one is more experienced this is perplexing because whilst there were only a few eyeing me from within, there was an angry mob without. I was pondering this problem when two most unwelcome guards climbed up my nostrils – fearing imminent death from suffocation I slammed the lid back on the hive and ran for home pursued by a diminishing number of bees; once there, I removed my veil, blew my nose and fortunately the bees shot out. I had only been stung in one nostril and on the head. Since then I only inspect the hives in fine warm weather when, like ourselves, they are more benign. My spaniel Drake, who also got stung now refuses to go near the bees in any weather.

At about this time a stranger reported that there was a swarm of bees in an alder tree by the river above the house. I checked my hives to make sure that the queens were at home and then telephoned my neighbour, whose swarm it was. He came down with an expert who climbed the tree and shook the swarm into a box. Unfortunately on the way down he slipped and the box and bees fell into the river. My neighbour, forgetting that he had not taken his veil, jumped in after them. Now swarming bees laden with honey for their new home may well be benevolent, but when suddenly immersed in cold water they become very angry indeed and my poor neighbour was badly stung. Many of the bees were drowned but the queen survived with sufficient followers to start again in a new palace.

This year Cleopatra, who succeeded Boadicea, swarmed whilst I was on holiday and flew with her entourage – the weather being hot – to The Black Horse, a delightful hostelry in our village. The landlord kindly telephone me but, there being no reply, Cleopatra continued on her journey to some unknown destination chosen for her by the scouts. She left behind a young queen,

Guinevere, who has few subjects but should do well next spring.

When I was going to take honey from Boadicea I used to put the clearer board beneath the top floor of the hive, the 'super', where the honey is stored. This cunning contrivance allows only one-way traffic down to the lower chambers. On one occasion, having lifted the supposedly bee-free but honey-full super from the hive, I was carrying it confidently homewards when, on reaching the footpath, I suddenly became aware that I was surrounded by angry bees – the super evidently had not fitted closely to the one below and the bees had had free access. As it was possible that about a dozen children on their ponies might come by and I had visions of a Thelwell-like stampede into the river, I took the box to the side of the house and smoked the bees away from as many frames as possible. By then the general alarm had been sounded and reinforcements arrived, so I decided to leave the last two or three frames and watched whilst the bees took back this honey to the hive from which it had so recently been removed. At least I got some honey.

When I was a boy the downland was a paradise for bees with a great variety of wild flowers, especially clovers. Sheep grazed along the slopes, the sheep bells of different notes blending in most melodious sound. Clover honey was delicious but now so many wild flowers are killed by spraying and so much of the downland of Hudson's *A Shepherd's Life* has vanished under the plough. In recent years great splashes of brilliant yellow startle the eye, as if from the brush of an abstract painter, on our landscape. Whilst the bees are certainly beneficial to the oil seed rape, the resulting honey has to my mind an unattractive taste and granulates too freely, although certainly excellent food for the bees themselves.

> *There's a whisper down the field, where the year has shot her yield*
> *and the ricks stand grey to the sun,*
> *Singing: 'Over then, come over, for the bee has quit the clover,*
> *and your English summer's done.'* (Kipling)

Nowadays, owing to change in agricultural practice, it would be more true to say that the clover has quit the bee. If Kipling lived today I feel sure that he would pen verses condemning the – to me as a non-farmer – horrific practice of stubble burning which darkens the summer skies, fills our homes and gardens with ash, and, if wantonly carried out, destroys so many hedges and helpless creatures, great and small.

Lately regulations and consciences have restored some improvement to our harvest scene.

## OTHER INSECTS

The mill, which is so much part of the landscape, is regarded by many insects as an ideal winter quarter for hibernation. In late autumn, as the nights become longer and colder, a great quantity and variety of flies fill the house, to Mrs Chitty's annoyance, and one autumn three years ago a large number of queen wasps appeared. We must have killed at least seventy. The following two years there were no wasp nests and very few wasps near the mill. I know that they are supposed to kill flies, and some indeed do, but I myself was delighted at their absence; delighted for my bees which had previously been robbed, and for my very few plums which were now not spoiled.

My most charming and gentle hibernators are the transparent lacewings, which particularly seem to like the mural on the staircase well where I try to persuade gullible guests that they are the lifelike and life-sized work of some distinguished artist – I seldom succeed!

Tortoiseshell butterflies also find the house comfortable for wintering if they can find a suitable curtain or cupboard, whilst innumerable spiders of varying sizes, shapes and manoeuvrability invade throughout the year. Their webs are a constant irritant to Mrs Chitty and me – one exceptionally crafty spider spins its web over the welcome light by the front door, so that luckless moths are attracted and greedily devoured. Mrs Chitty is too kind-hearted to kill them and when one falls into my bath, I allow it to take its chance down the plug hole!

# DRAKE

My black and white spaniel Drake is a delightful companion, except out shooting for which he was trained. For a successful gundog there are two essentials, obedience and a good handler, and as I have owned four dogs in my life all of which barked and 'ran in' I can hardly claim to be the latter.

At a pheasant shoot he is a disaster as he will yell in frustration throughout the drive and, as if this fault were not enough, will sometimes vanish into the covert which is about to be driven if he considers that he has not had enough exercise. He is therefore continually being shown yellow and red cards, sent off the field and suspended. However he retrieves, marks well and is silent when pigeon shooting – perhaps when he is older he will reform but so far he shows no such improvement. Fishing he finds most boring and gardening extremely dull.

On the whole he is well-behaved in the house though he has recently developed the tiresome habit of springing onto my or my guests' laps at the moment when they are enjoying a cup of coffee or glass of port, or reading a newspaper. A week or so ago he leapt through the service hatch from the dining-room to the kitchen, narrowly missing the breakfast tray coming the other way and endeavouring to snatch a sausage as he flew past. If he wants to go out in the night he brings me one slipper, whilst the second slipper denotes extreme urgency. One moonlight night after such a red alert I let him out and watched him go to the river and drink, returning at once. There was a bowl of water in my bedroom and another in the hall and far from being horrible chlorinated water it comes from a bore hole!

At Drake's suggestion we decided to hold a field trial for the Worst Dog in Wiltshire award. Invitations were sent to a varied collection of likely candidates.

On the appointed day the first arrival was Drake's nearest neighbour and girlfriend, a black Labrador bitch named Spindle, a hot favourite for the title. She went into an early lead, to the delight of her backers, when the judges awarded her minus ten

points on the arrival of her rather breathless owner, after an absence of some half an hour, apologizing for not bringing Spindle as she could not find her.

The first test was a water trial just above the mill. Seven dogs were chosen for the first heat: two normally steady black labradors belonging to one of the judges; Spindle; Tups, a yellow dog of character but doubtful parentage; two other yellows and my black and white spaniel, Drake. Each was given the starting order. The judge threw the dummy into the river before most handlers were ready, so that all the contestants including his own 'steadies' leapt simultaneously into the river together with several from the next heat. The resulting scrimmage for possession churned the water into foam and resembled some strange canine water polo match – yellows v. blacks with a black and white referee. Spindle eventually seized the dummy, landed with it on the far bank, and returned with it via the bridge before depositing it at her owner's feet whilst shaking the water and weeds all over her handler. She was awarded minus fifteen points. By this time the judges, and the trout, were in a state of complete bewilderment.

The final trial was a test of retrieving the dummy fired from a sort of rocket launcher but smaller than those used at Cape Canaveral. The dogs were supposed to wait until called by the judges, but once again the majority broke in every direction. Those which were gun-shy went backwards – minus five points. The dummy soared in a perfect parabola far into the meadow, landing nose first in a cowpat. Spindle, on being ordered to fetch it, was overheard by Drake to say 'Fetch it yourself' (minus twenty points). Drake lost any hope of a place by picking the dummy up by the end and returning it sticking out like a cigar smoked by some canine tycoon, but did score minus five points for refusing to drop it when told to do so. Spindle was an easy winner of the First Prize for the Worst Dog in Wiltshire award whilst the beautifully bred and impeccably behaved Meg won the minor award for the best behaved. Dipper, an enthusiastic labrador, and Spindle wrote delightful letters of thanks to Drake.

# The River

*I chatter chatter as I flow, to join the brimming river*
*For men may come and men may go, but I go on for ever*
                                                    (Tennyson)

The river as it flows through and round the house can be seen in three different aspects. Above, the channel which was dug by hand – an amazing task when you think of the mechanical juggernauts used today – narrows to a 'V' as it approaches the house, so that the water is forced through the mill, formerly to drive the sails of the mill wheel and thus turn the grinding stones. The water held up there is deeper and runs more slowly before tumbling through the hatches into the mill pool itself (if not diverted through the house), the amount of flow being regulated by the hatches of the mill pool or an undershot hatch within the mill.

One of the problems is control of weeds, especially at weed-cutting time, and at first I had no weed rack. One Monday morning a huge weed jam had built up as the river tried to force the weeds under the hatch through the narrowing 'V', the compressed weeds had backed up some fifty yards, and scarcely any water was flowing through the house. As there was an increasing stench of rotting vegetation, I decided at dawn to try to clear the weeds before catching my train to London. So I put on my thigh boots and, armed with a slasher, lowered myself into the river with high walls on both sides, to be confronted by a solid wall of weeds jammed tight below the hatch like a coal face. After some futile slashing, I decided to put the blade into the weed jam horizontally and to cut across. This ploy was surprisingly success-ful as suddenly the dam began to bulge, water seeped through with ever increasing force and, as I seized the rail above the hatch,

49

the dam burst with a roar and my feet were engulfed in tons of weed. My main fear was that I would be swept under the weed and suffocated, but I decided that I could not hang on any longer, so let go and shot out of my house feet first at some ten knots, touching down thankfully in the shallows some twenty yards below. As the water was coming from beneath the hatch, I was propelled upwards so there was in fact little or no danger of being engulfed by weed; there was however a real danger of banging my head against the sides as I was carried through the narrow stone-lined channel before emerging covered in mud and weed, but very ready for breakfast. I envied Ophelia's flower-bedecked journey but not her fate.

In those days our early train had a breakfast car and four of us used to gather there. On this particular morning all went well as far as Woking when we learned that a signal box had almost fallen on the line and that trains could progress no further. One of our number, a solicitor, very sensibly dashed across the platform and caught the next train home, whilst my other companions, a publisher and a book expert from Sotheby's, accepted a lift from a kind stranger whose car was at Woking. This proved to be a Jaguar which, since he was late for an appointment, he drove at great speed. Being rather superstitious I believe that disasters are often threefold, so I expected a fatal crash at every crossroads; our good Samaritan dropped us off safely, however. Even for a Monday, this was a most unsatisfactory start to the day.

This brings me to the southern aspect of the mill, and the stream which my picture window overlooks, where I spend so much time now that I have retired when I should be gardening, or more usefully engaged. Here the stream is narrow and shallow.

To the west lies the mill pool which as I write is in flood – an impressive sight. Yesterday the hatches had jammed but my heavy nephew and very light great-nephews arrived in the nick of time. Whilst the latter went paddling amongst the daffodils in my flooded garden, instantly filling their boots, my nephew and I managed to open one hatch, so now the water level has dropped and my garden is re-emerging – perhaps in time to save the young primulas from drowning. The excitement of flood time must be even greater to the fishy inhabitants below the surface than it is to those of us who live above the waters. Soon the kingfisher will return to his perch but will he bring his mate to inspect our refurbished nesting bank?

Of the fish which inhabit the river, brown trout must take pride of place. Our native fish were golden hued with a few distinctive red spots but now, through introductions from fish farms – essential to allow more fish for the increased number of fishermen who enjoy our peaceful valley – the trout are often more speckled. The Piscatorial Society hopes that eventually, by careful management, the stock of brown trout will be maintained without any introductions. Rainbow trout, which breed naturally in only a few rivers, are not encouraged by the Society however good they may be for sport and the kitchen.

Large trout, and once a very large grayling, often lie above the mill hatch and several rise along the weed rack where the current

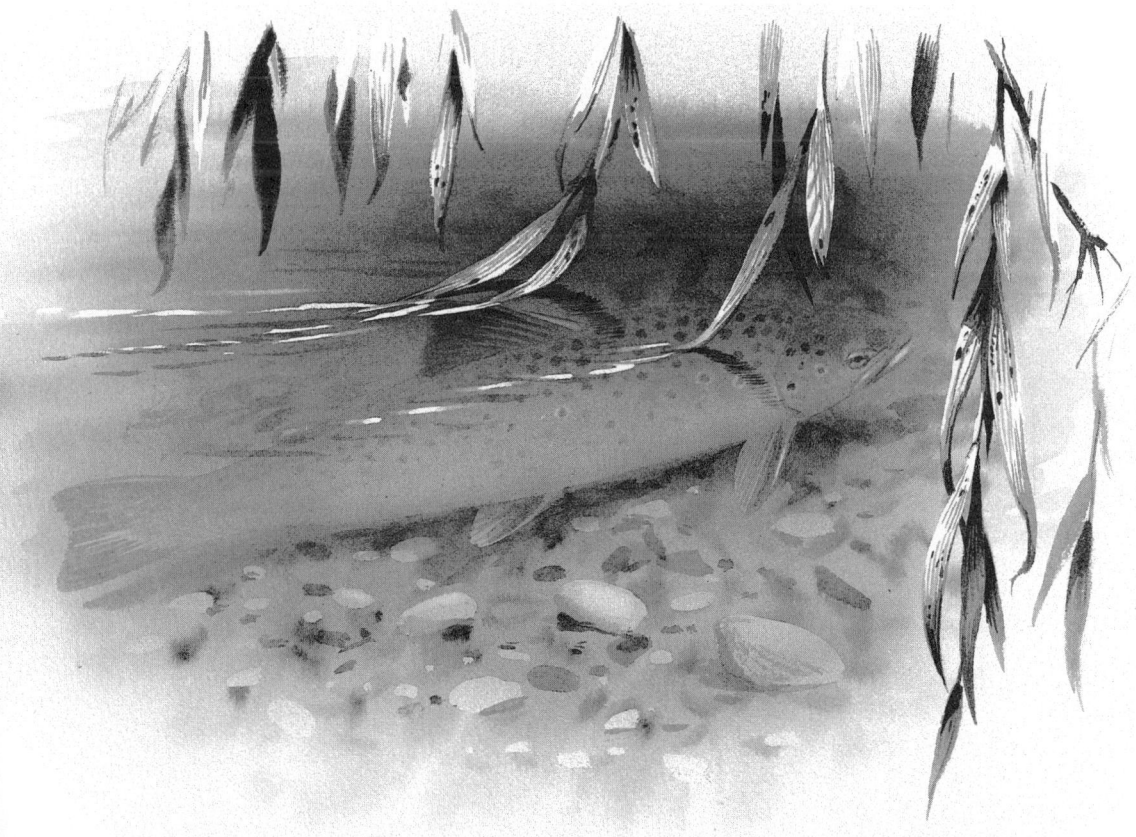

serves up an endless supply of supper especially during mayfly time. Last May, in very wet weather rafts of living flies, spent gnats and the shucks from which they had first emerged drifted continuously past to be engulfed by the yawning hatches into the mill pool's voracious maw. In wet and very windy weather the flies are unable to take off and fly to nearby reeds and boughs, and thus perish. The largest trout I saw appeared just below the weed rack on the day following the season's end. He had his wife, or perhaps fiancée, with him so I wished them well.

The trout below the mill in the pool through which I once sped are tame, as I feed them, and these are the preserve of my great-nephew who as yet is seldom successful in catching them. His future as a fisherman was assured in my mind when my nephew, having spent an entire morning trying to teach his son how to cast, finally left him in exasperation. The boy then came to me and said, 'The trouble with Dad is he is too impatient'!

One Sunday morning I woke early thinking that I heard muffled voices below my window, peered out, and saw four sinister men in camouflaged jackets hauling out my precious trout with hand lines. I was so incensed that I screamed at them. They dropped everything and fled on to the island. I put on a sweater and trousers and ran down to the little bridge which spans the outflow, feeling like a very ill-protected Horatius. Eventually they emerged from the bushes and waded across the main river leaving behind spools of nylon, several hooks and a chair leg which fortunately they did not use. The trout had been removed whilst I dressed. These were not local men but professional poachers and probably the same gang that snatch helpless salmon on their spawning redds, at the very culmination of their epic travels to reproduce their kind.

I have every sympathy for the lads of the village with fishing aspirations – 'the bread brigade' as I call them. I usually make them junior members of the excellent Salisbury Anglers Club, where several have now graduated to successful fly fishermen in summer and to catchers of ill-named 'coarse' fish in winter.

The beautiful silvery grayling need no encouragement and

survive in such numbers that some are removed each year by nets or electric fishing and transported to waters where they are scarce. They prefer to swim in shoals which is perhaps the reason for their survival, as is the case with many flocks of birds. The French term for grayling is 'ombre', so apt since it is often the 'shadow' alone which reveals the presence of these fish lying, as they usually prefer, near the bottom, until their rises from the depths for flies leave tell-tale bubbles.

Once in early September 1981 I opened the door of my river room and there was the tail of a salmon wagging to keep station above the hatch. I peered cautiously over the hatch and guessed the weight at about 12 pounds. This fish and an osprey have been my most exciting visitors.

I have always been interested in the mysterious migrations of salmon; not only at sea but within their native rivers. I fish each year, through the great kindness of friends, on the Spey and on this beautiful beat lies a pool, surmounted by a steep bank of larch and birches, which is idyllic on a summer's evening when graceful roe in their red summer coats come down to drink and to feed at the pool's edge. The water on this long pool is swift and shallow flowing into a deep hole beneath a rock on which kestrels nest. Here lie the salmon at summer's height but as the light begins to fade so the fish, both salmon and sea trout, begin to show at the tail and plop their way up to the neck and beyond. I know not how far they go, returning before dawn to their rocky fastness.

But I have strayed too long to this enchanted pool and must return to my salmon. Now, I thought, I will be able to monitor her comings and goings but alas, weed cutting was in progress and she so disliked constant weed flowing into her gills that she was gone the next day, returning briefly on two occasions.

It is only in recent years that salmon have spawned on this part of the Avon, due to the efforts of the Wessex Water Authority in providing salmon ladders to bypass hatches and difficult weirs. One such ladder may be seen in the heart of Salisbury where several pairs attempt to spawn each year at their peril, were it not for the dedicated vigilance of our local water bailiff.

Of the other denizens of the river, the voracious pike for which we used to spin in winter are now controlled by electric fishing whilst roach, rudd and dace are transferred to other waters along with some grayling – these abound and provide fishing for several enthusiasts in the cold early winter months while trout are left in peace to spawn. Eels, too, are removed although many live for years before their return migration to sea en route for the Sargasso spawning grounds which, experts now believe, many never reach.

Until a few days ago crayfish, the delicious little freshwater lobsters, abounded under the stones and banks. Each summer we had a barbecue here and when the light faded, circular nets baited with fish or meat were lowered into the river at strategic points. There they should remain for at least twenty minutes to allow crayfish to gather for the feast. In fact children, for whom the party was really given, found that an inspection of the nets every few minutes was essential; torches were flashed and the crayfish returned to their lairs. The few that were caught were usually undersized and returned gleefully to the river. By the end of the evening, every child and most adults were soaking wet, covered in mud but blissfully happy and the memory of flashing torch lights and merry laughter lingered on for several nights. On one occasion a short-sighted father walked straight into the river with a splendid splash, to the delight of his own and other children, who learnt several strange oaths which no doubt they found useful on future fishy expeditions, when faced with similar disasters.

And now a real tragedy has struck. The dreaded crayfish plague has killed virtually all our crayfish, almost overnight. I could find none at all in the shallows recently and my great-nephew, an ardent and more skilled crayfisher than I, could find but one very small crayfishlet. Those happy crayfishing nights, those flashing torches, the children's excited screams and laughter are indeed now only a memory – perhaps for ever.

## DRY FLY FISHING

The charm of dry fly fishing has been best described by Lord Grey in his delightful book, *Fly Fishing*. I have mentioned elsewhere the heron, with more unorthodox methods of fishing, which I named after him. After his second marriage Lord Grey did indeed live in this valley, only a couple of miles upstream, but, sadly, failing eyesight robbed him of the pleasure of his favourite sport.

My father used to love to quote Lord Grey's reply to someone who was rather scornful of the sport, to this effect: 'If you don't like fishing, don't fish – why should you? But . . . tell me something that you like as much as I like fishing!' I have failed to find this quotation in *Fly Fishing* so perhaps my father heard him say this when they were fellow members of a syndicate at Itchen Abbas, where I believe that Lord Grey's fishing cottage was accidentally burnt, together with his fishing records – a grievous loss.

Here the fishing is mostly dry fly, which to its devotees is the best of all fishing, though nymphs are sometimes used too. We do not get the early grannom which hatch in such profusion on some rivers and indeed our season does not start until May, although fishermen begin to become restless long before and may be seen haunting fishing tackle shops, leaning over bridges, pacing the banks of rivers and gathering in restful havens like the Flyfishers' Club and various fishing haunts to discuss with optimism the season's prospects. At the beginning of May there should be plenty of fly about at midday but except on warm evenings not much fall of spinners – the females of the species dying on the water after fulfilling their short life's purpose of laying eggs. The dance of the males over the meadows (like puppets on invisible strings) as the setting sun illuminates them and paints the willows and reeds in golden hues, is a magical time for the waiting fishermen – especially at the end of May. As the season advances evenings provide the best sport, and September is probably the best fly month though the fish are then more difficult to catch.

It is often said that the nearer you live to a river the less you fish and I confess that I find this to be true. After all, the river and the trout are always here, part of the scenery, and there may be letters to write or work to be done in the garden, where weeds grow with the pace of lost monsters.

Away from home an invitation to the Test or Itchen is an exciting day's outing, everything to be assembled the night before, whereas here I wander out when the weather and conditions are favourable. The stretch of river above the house holds many trout and I have only to look out of the window to see if the rise has started. How fortunate I am!

As the season advances I will have probably located many of the larger fish which are to be found each year in certain favoured places – where the flow of current brings the most flies, or where an overhanging alder provides protection, shade and fallen delicacies. I have never been cured of my habit of believing that a larger fish is certain to be rising just round the next bend. These fish almost become friends and I am sure that they recognize me, as I do them, and derive considerable satisfaction from continuing to rise at every fly except mine. Others flee at my approach or as my first cast either glints in the sun or lands too heavily in the wrong place.

The planning of a day's or even an evening's fishing is important. The bank where the kingfisher nests is overhung with beautiful, dense yew trees of considerable age. The current flows on the far side, so there the fish lie, and sometimes cruise where the current is slacker. Drag, one of our most irritating problems but the fish's friend, draws the fly away at the very moment when

surely the fish would take. Here too the light fails more swiftly and it becomes increasingly difficult to discern whether the rise is at one's own fly, so I must drop downstream where there are fewer trees and where the twilight in the west lingers longer. I have fished unsuccessfully for certain trout for an entire season. Each time there is a thrill at the sight of a heavy rise under the bank. The fly is changed, an operation which becomes most difficult as darkness falls, so I usually carry a spare cast which will be more easily seen by fishermen and perhaps by fish than the Sherry spinner, Lunn's Particular or whatever offering I need to replace.

The evening rise is very special and the valley is hushed as the shadows lengthen; the rises are seen and not heard. Ducks may be flighting out to the barley fields and swifts scream overhead before soaring to sleep on the wing (or so experts believe; surely though they are still awake and possibly even still feeding, like the water bats which replace the swallow families at dusk). Whatever the outcome we return home mentally refreshed, if at times frustrated; hungry for supper, even if it is again bacon and eggs; and full of hard luck stories which no one really believes – nor are they expected to do so.

By day there is no rush and I prefer the shallows where fish can be seen nymphing amongst the weeds or taking the occasional fly with leisured accuracy. There is a great excitement at watching a feeding fish, seeing him inspect the fly or perhaps dash for cover. The larger the fish, the slower should be the strike or tightening of the line. Alas! I find that when I cast for a big fish, it usually rises so slowly to my fly that in my excitement I strike too soon and the fish and my supper have gone, with a great swirl of water which adds even more weight to my hard luck story.

The evening rise is all too short, especially as autumn advances, and speed is disastrous when fishing. Therein perhaps may lie one of fishing's greatest charms in an age of continual rush and noise. As the wise Walton wrote, 'Study to be quiet.'

The fishing here has changed in many ways since I was a boy and caught my first grayling – so small that I cast again with it, thinking that it was a piece of weed. At the second attempt the little fish flew over my head and was only found after a considerable search in the long grass and carried home in triumph. All the fish were wild and whilst there were fewer than today, they were larger, a few weighing three pounds or more. This was partly due to less competition for food which in turn was due to the numbers of pike, some of ten pounds or more, for which we used to spin in the winter. My father was an ardent pike fisherman and in his younger days also spent much time fishing for the elusive Thames trout.

Now the river is carefully stocked and much as one deplores the lose of truly wild fishing, the river provides many members of the excellent Piscatorial Society with good fishing for both trout and grayling and many with happy evenings amongst friends. The Society makes every effort to encourage wild fish to spawn and multiply.

Due to water abstraction, the average flow is far less and the site of the boat house from which we used to paddle our canoe or punt has long since been high and dry. One of our favourite sports in those distant days was to spear eels from a canoe. An eel spear, which is much used in the fen country of East Anglia, consists of three or four prongs widening at the end with barbs at intervals between them. The eel is not transfixed by the points of the spear which are relatively blunt. The spear prongs straddle the eel which is then drawn out of the water attached to the barbs. This is the theory. In practice the result was very different, as the following tale will show.

I was the spearman, standing in the bows and my brother was paddling. He was supposed to be keeping at least one eye on the far bank in case anyone might be fishing for trout. Suddenly I saw an eel rather too far out off the starboard bow. We were travelling too fast to alter course quickly so I lunged at the eel with the long-handled spear. The eel was lying on the gravel patch so that the prongs merely hit the hard bottom, the canoe capsized and I went a long way down. I was wearing a cap to shade my eyes and as I surfaced I saw that a fisherman had been about to cast for a fish just where we had overturned. I doffed my cap to him, spluttered an apology and swam off downstream after my brother and the up-turned canoe. When I looked back the fisherman was still kneeling in the same position, as if he himself had been transfixed. The eel was no doubt amused.

I must conlude this river section with a few verses from Tennyson's 'The Brook', replying to 'O babbling brook whence come you?', as they seem so appropriate to this mill and its garden:

*I come from haunts of coot and hern*
*I make a sudden sally,*
*and sparkle out among the fern*
*to bicker down a valley.*

*I wind about and in and out*
*with here a blossom sailing*
*And here and there a lusty trout*
*and here and there a grayling.*
*And here and there a foamy flake*
*upon me, as I travel*
*With many a silver waterbreak*
*Above the golden gravel.*

*And draw them all along, and flow*
*to join the brimming river*
*For men may come and men may go*
*but I go on for ever.*

Yes! In my brief tenure of this beautiful mill, I and my successors must always remember those last two lines, and endeavour to care for our river and its many inhabitants.

## WATER MEADOWS

*Meadows trim with daisies pied*
*Shallow brooks and rivers wide.*

(Milton)

Before the war, the water meadows were flooded (or 'drowned' as we call it) each winter by a series of channels which had to be dug out with special tools. The water was diverted by hatches into a carrier and then flowed evenly across the meadow before returning to the river itself. This irrigation was most beneficial, kept off the frost and thus produced an early bite of grass for the cattle. The man in charge was called by the rather sinister name of 'Drowner'; ours had a mongrel dog which used to catch the moles which collected to feed on the worms. Sometimes the meadows were drowned again in a dry summer. Today, some meadows further downstream are still flooded, but with the aid of a sort of mechanical digger.

This water meadow system was of benefit not only to the cattle but indirectly to the river since much food was washed in for the trout. In hard winters many birds came to feed, as snipe could probe with their soft bills and lapwings, those delightful birds which tumble and call over their nesting downs and fields throughout the summer, found food denied to them elsewhere by the frozen ground. My brother and I used to try to shoot the snipe – very difficult with their zig-zag flight but delicious to eat when we succeeded.

My own solitary meadow is a delight in the spring with an abundance of the beautiful marsh orchids, yellow flags with islands of kingcups amongst the reeds and rushes and the graceful little water avens. The farmer whose cows graze there readily agreed to forgo any spraying, so they continue to flourish.

# The Birds

My picture window faces sou-west and looks straight down the river flanked by willows and rushes, whilst the mill itself forms an effective screen from disturbance from people, ponies and dogs passing along the bridle path. I use the table in front of the window for breakfast, for writing or just for watching, always with my binoculars at my elbow. Writing or even eating is constantly interrupted to check on passing birds or movement in rushes or river. Robert Dougall remarked that it was the only bird hide which he knew from which you could watch birds whilst drinking a gin and tonic!

## THE KINGFISHER

The kingfisher must take pride of place – or, more correctly, perch. This year's bird is a cock named Halcyon. Every morning when I draw back my curtains I look to see if he is on the perch which I erected for him, or on the willow beside the perch. Except in the breeding season he is there as often as not, a living jewel if the sun is shining and the focal point for all birdwatching activities or photography. He returns again in the evenings. I now know that the reason for this is that 'miller's thumbs' emerge from beneath their stones at dusk when it is safer to start their nocturnal forays. The pool below the house is a perfect residential area for them because the bottom is strewn with brick and stone rubble from the old mill.

Rodger McPhail, the illustrator of this book and the most excellent observer and rapid sketcher from life, has just been staying. Halcyon has been fishing each morning for the past fortnight and is here more continuously than I can remember. His voracity and indeed capacity is astonishing. He caught three 'miller's thumbs' whilst we ate our eggs and bacon and shortly after caught a very large one which he carried to a branch a few feet from the shore. A fish of this size may require 'bonking' for some five minutes and on this occasion he dropped it and walked round it on the ground to devour it – a rare sight! Another he dropped from his perch but this he quickly retrieved from the river.

My great-nephew and -niece arrived for lunch and spent much of the afternoon looking to see if any crayfish had survived the plague, turning over many stones and disturbing many 'miller's thumbs' and a few stone loach. When they came in for tea, so did Halycon the king of the fishers return for his evening repast and thus found an unexpected abundance of fish. We guessed that we had seen him eat over a dozen fish during the day and no doubt he fished elsewhere and caught fish here whilst we were not watching! We also calculated that Halcyon caught about one fish in every three dives. Once he hovered for several seconds before deciding that his prospective dinner was in too deep water or perhaps, unlike human anglers, was too large to catch!

Many crayfish too lived there before the plague struck but, presumably because of their hard shells and claws, none except the trout, and sometimes myself, ate them. 'Miller's thumbs' do resemble a thumb which has been hit by a hammer, with a rather thin tail attached and a spiny back fin which cause the kingfishers to 'bonk' the poor fish on some branch several times before swallowing it. Both water rails, which probe beneath the stones and so are not confined to dawn and dusk, and the ever-questing dabchicks also find these little fish plentiful and delicious.

I refer to 'him' because this year's bird, Halcyon, is a male: his lower mandible is black whereas 'hers' is pink. Kingfishers are very territorial so, except for a brief period in the spring, I never see two together nor, sadly, have they ever brought their young to my perch. There is no question of another kingfisher taking part in the fishing feast. How can such a comparatively small bird have the capacity and digestive ability to devour perhaps its own weight of fish in a single day? But then only last week I saw a heron catch a trout of over 3 pounds and then continue to fish.

Ron and Rosemary Eastman, who made the wonderful film on the life of the kingfisher and produced a fascinating book on the subject, kindly came over to lunch to tell me how to construct a nesting bank. This we did, and even planted daffodils and snowdrops for their roof garden. The Eastmans told me not to put it too near the house, so we placed it beside their perch, about 25 yards away. Our soil is rather chalky and soon began to crumble, so the birds have never used it. I hope to restore it next week and perhaps build another further away but there is no point if it cannot be seen from the window. At present they nest, I believe, on the banks of a stream overhung with yew boughs about half a mile above the house.

My efforts at filming the bird fishing are pathetic compared with the Eastmans'. I have some good moments of catching fish but the bird then usually takes these to a different branch, and although my camera will zoom quickly, it goes out of focus and has to be adjusted by hand. Once, after a successful dive filmed in slow motion, the kingfisher brought his large 'miller's thumb' to a branch quite close. I forgot to vary the focus in my excitement, so the result was more like Jack Nicklaus in one of his rare visits to a bunker. Our English kingfishers, unlike most of the other species, seldom hover – although they occasionally do so just before entering the water – but I saw one hover above a mayfly which it caught as it took off for its maiden flight. They seem to prefer fishing from about 6 to 20 feet above the water and to keep to shallow water of a depth of less than two feet.

I did have one most fortunate success whilst filming. The hen kingfisher who alone used my perch that year – she had a white ruff behind the head, which is not typical – was busy fishing from her perch when I saw her mate flying up the river. He settled on her and they mated, and I even remembered to zoom in, though I felt I was 'prying'. Later I read that the male should present a fish to his about-to-be bride as a charming gesture or wedding breakfast. My ill-mannered bird made no such offer. I told this to a girl-friend who replied, 'I am very fond of smoked salmon.' Alas! I had none.

Kingfishers seem to be able to see or in some way sense the movement of fish as far as twenty yards away and I have seen Halcyon catch a fish when the light has almost gone. I once tried to see if he would eat a sardine and so dropped one out of my bedroom window as he sat on the perch. It sank like a stone and an oil slick rose. I did not repeat the experiment as I am very fond of sardines. I wonder if they would have been a substitute for smoked salmon?

We have today replaced with clay, borrowed from a layer a foot below Isabella's Orchard, the gaps where the soil has crumbled from their nesting bank, and patched up the turf on their roof garden lawn. The result at present looks irresistible but I am wondering whether it is the cock or the hen that chooses the site. I think probably it is the hen, so unless the cock can persuade her to nest here when she joins him soon, they will go elsewhere, to her chosen site, perhaps.

> '. . . a bank where the wild thyme blows
> whereon oxlips and the nodding violet grows . . .'

But how could they spurn my daffodil- and snowdrop-mantled bank? We shall see.

# THE WATER RAIL

Today, the water rail has been hurrying from stone to stone like a housewife searching for the best bargains at the fish market. Perhaps he (or she) is building up his reserves of strength for the long journey northwards in a month's time. The kingfisher too has been fishing from willows further down the stream but his migration, if go he must, will only be a few hundred yards north to the seclusion of the yew-shaded little stream where his parents nested, or as near as they will allow. Water rails feed furtively, mostly along the reeds at the water's edge, occasionally venturing in their nervous manner to seek worms further inland, or wading out into the shallows, probing beneath the stones and hurrying ashore with any catch, like salmon fishermen with a fish on the gaff.

Before I came to live here, I believed that water rails were with us always, and that I did not see them in spring or summer because there was then so much cover into which these shy birds could disappear. But, from this window which is the focal point of this book, the water rails in winter and the water voles in summer are continually crossing the stream like ferryboats, the water rails with their jerky movements as if propelled by unseen oars, their heads thrust forward in time with the stroke like that of a cox in a boat race. The thought eventually occurred to me that they would be equally visible in summer or winter as they crossed, as one has just done! So I referred to my *Birds of Wiltshire* by the Reverend A. G. Smith, which I greatly admire especially for the great bustard cock which adorns the cover. I was however sorry to read that he used to shoot water rails, as did many of his contemporaries. He did not realize that these rails migrate; perhaps not all did in his day, but a modern *Birds of Wiltshire* has no record of their breeding in our county, and Dillon Ripley's recent book *The Rails* suggests that most British water rails go to Iceland, Norway or Sweden to breed. They do however also frequently nest in the Broadland district of East Anglia.

I have already described their fishing habits and I read that they have been known to kill small birds in hard times so perhaps it is as well that these beautiful and elegant birds do leave us in the spring. I am very loathe, though, to believe that my gentle friends have such habits! When they return around the first week of November the resident moorhens chase them continually but after a while they are left to feed in peace. Why our rails should migrate when there is so much food for their young in the reed beds is a mystery but at least they are more common than ever in the winter and have not suffered the fate of the land rails through the change of farming operations. When I was a boy, these equally delightful little birds and a few quail, if never common, were seen each year at harvest time.

# DABCHICKS, 'POP-UPS' OR
# LITTLE GREBES

These delightful and amusing little birds are one of my favourites and one of those species which my window allows me to watch at close quarters. Thus I am frequently able to observe them feeding with the mallard, diving beneath them as the ducks pull up the water weeds and thus flush water shrimps and little fish from their hiding places. They are tireless feeders and like most birds never seem to be satisfied. I have often watched them catch a large fish which they manage to kill by immersing it – letting go and then catching it again, thus presumably exhausting and eventually drowning it. Their food just below the mill is mostly 'miller's thumbs' because these are the most plentiful species and they can fish where the kingfisher, limited by perch and depth, and the water rail also by depth, are unable to reach, but of course they poach the shallows of the other two species' preserves. Dabchicks, and to a lesser extent water rails, appear to be incapable of flight as they flutter across the surface, the rails with their legs trailing, but both are partial migrants and on one occasion I saw six dabchicks on the lake in St James's Park; they had evidently flown in during the night. They nest each year in the rushes above the mill – their floating nest is difficult to see as the hens cover the eggs rapidly as you approach. I have filmed the parents feeding the greedy offspring with shrimps and various small water nymphs and fry and when the young are taught to fish for themselves, fish are released and re-caught by the parents until they become more adept.

When I myself am fishing I often see an apparently good rise under the bank. I stalk this and hover, ready to cast, when the rise is repeated a little higher up. Finally the dabchick dives in alarm, splashing the surface, and I often think that I can hear a chortle of amusement come to the surface with the bubbles. Whilst normally of a peaceful disposition, dabchicks do occasionally annoy one another and a chase will ensue above and below the surface with many changes of direction until the pursued evades the pursuer or more probably they both tire of the quarrel, like human beings.

## MOORHENS AND COOTS

Moorhens and, more appropriately, water hens are ever present on the river or lawn. In spring they spread their tails so that the white shows, almost in the manner of blackcock on their lek. Moorhens are quarrelsome and territorial at nesting time and will sometimes attack small ducklings. They are however invariably driven off by the mother duck, and soon peaceful relations are resumed. After the young have hatched the parents will often build a new and very temporary nest a yard or so out from the bank as a night nursery, away from any prowling grey rats. Water voles are strictly vegetarian, so do not pose a threat. Unlike mallard, moorhens are diurnal and will often roost in willow trees above the water where they are safer from prowling foxes, cats or the like.

Last summer a large hen sparrowhawk settled beside the little side stream which runs through my garden, presumably to drink and bathe. A moorhen a few yards away immediately flew at her and she retreated in disarray in a most cowardly fashion. A moorhen when puffed out with head held low is a formidable opponent. I was reminded of the time when I saw about a dozen rooks pull the tail of one of our cock great bustards at the feeding hopper; he retired without too much loss of dignity, and then the rooks pulled the tail of a cock pheasant who in their view had eaten enough. He too retired. Then a pair of grey partridges arrived, surveyed the scene and, lowering their heads and puffing out their feathers like the moorhen, charged the rooks all of whom fell back and dared not approach again until the partridges had finished their evening meal!

A pair of quarrelsome coots have returned to the stream and have already stuck out their bald heads and necks at water level and steamed at full speed ahead to threaten the moorhens with a feint ramming. This churning of the placid waters will now continue until nesting territories have been established and, indeed, until the rather hideous little offspring are fully fledged. The spring offensive is with us: mallard drakes chase each other; both chase the ducks; coots chase moorhens; moorhens chase coots; and dabchicks chase each other as they dive beneath the contestants with loud cries from all, but few injuries.

## WILDFOWL

Wildfowl have always fascinated me. True wildfowling is a most exciting sport and too many of my Cambridge days and nights were spent on the Wash with my friends when we should have been studying. Perhaps my wise tutor realized that wildfowling at least taught endurance and the study of nature in all her moods, since he often cooked breakfast for us in the very early hours before our precarious drive to the coast in order to arrive in time for the morning flight. The call of geese increasing with the light as they prepared to fly inland and the whistling of widgeon, curlew and the many varieties of waders moving with the ebb or flow of the tide, are a lasting memory.

Here, mallard are by far the commonest species of duck. They are mostly residents and the majority nest in trees. The pollarded willows afford most desirable nesting places, safe from the floods which sometimes occur in spring when the growth of weed is heavy, and from the foxes, rats and stoats that may hunt the banks. Some years ago, I bought three wicker nesting baskets for ducks and put them in various willow trees. My elder brother was somewhat scornful of their potential so we had a bet of twelve bottles of claret to one, that none would nest therein. (I lost for the first two springs but took care to give him a bottle of a good vintage on each occasion.) The third year ducks nested in each of the three baskets, and whilst I did not press for three dozen, I did claim and received the 'Connoisseur's Dozen', as listed in some catalogue.

Albino or partially albino birds are most useful in assessing territory or age. A white, probably albino, duck has nested for two years at the end of the island. Her first brood of eight remained with her until the winter. The second year she lost her first nest, probably from floods since she always nests on the ground, but her second brood of four survived. Mrs Chitty, who likes to name all birds, calls her Dilly. There is another called Dally who keeps company with some twenty other mallard at the

top of the field above the house and has also been here for three years; although she and her companions fly down to the shallows below the island she nests somewhere above the mill.

In early March Dilly appeared near her old nesting site followed at first by two quarrelling suitors but later by only one – I hope her true husband of yesteryear. Promiscuous would be a more apt description than polygamous of drake mallard. The ducks are chased unmercifully by two, three or more drakes in aerial pursuit and then on the water. I have seen three randy drakes attempt to rape a duck so that I feared she would drown as they held her beneath the surface. Eventually she took to the land, evidently deciding that 'a fate worse than death' was infinitely to be preferred to death by drowning. As I write Dilly has just reappeared with her mate. The rival is a few yards below with attendant dabchick.

By mid-April Dilly had been sitting for some time so I walked quietly down towards the end of the island, the site of her previous nest. I am not clever at finding nests, but even I could see Dilly perched high on hers – as ill-concealed and as obvious as a tennis ball on a golf tee.

Appropriately on Easter morning Dilly's babies hatched and I watched her lead her splendid brood of a dozen ducklings down the bank for their first swim. There was no white feather in colour or character. Each night she returned with them to their duvet of soft down.

Dilly's '84 brood of ducklings proved to be more adventurous than usual and certainly more disobedient: they often scattered themselves widely across the mill pool, a frequent cause of separation and subsequent loss. Eventually, when only three half-grown ducklings remained, she was rejoined by her husband. The next day the ducklings were gone, having I presumed been told in modern parlance to 'get lost'. Evidently one was put in charge, as a couple of days later I saw what I supposed was their flotilla steaming in line down the centre of the river towards the treacherous mill hatches. However the leader climbed through the weed rack, called to the other two – who were busy catching mayflies – to follow and conducted them safely down the shallow overflow sluice to the mill pool and thence to the comparative safety of the weed beds in the shallows below. We now wait to see whether Dilly is nesting for a second time.

Some of our local mallard move at times down to the flooded meadows of the lower Avon, some perhaps as far as the coast in search of food, but I believe that we have within a mile, from the willow bed above to Broad Water below, a more or less resident population of between fifty and a hundred mallard with only occasional visitors. In summer they fly to the harvest fields at dusk and again before dawn.

Widgeon, alas, are seldom seen. When they are here, usually in hard weather, they prefer the wide shallow water at the very limit of my vision from the window, about four hundred yards downstream.

Graceful teal, beautiful in plumage and flight, come each autumn but seldom stay near the mill, seeming to prefer the muddy back streams to the main river. Of the diving ducks, tufted duck nest at times and are often present in winter on the deeper water above the house and a very fine drake pochard once remained in solitary state, becoming so tame that he flew only a short way, if at all, as I walked my dog, Drake, past him. This winter a dozen Gadwall duck arrived about a mile further upstream.

## SWANS

Mute swans are the largest and most majestic of our river birds. About a hundred Bewick swans come each winter to the Lower Avon together with whitefronted geese, but I have neither seen nor heard either species in our upper, narrower valley. A pair of mute swans nest on the island each year and I particularly like to watch the newly hatched young when they climb on board their parents' backs for a trip up or down river – sometimes only their heads showing above the wing coverts.

An old cob driving away a rival, or even his puzzled offspring of the previous year, is an awesome sight. Sometimes as he comes down the river with wings beating the water, neck arched with beak extended, he will corner them above the house, delivering sharp pecks as they try to escape. I have seen them flee through the hatches and once through the house itself. I ran to the lower window to watch the fugitives appear dishevelled but safe below. Such is the fury of territorial aggression.

By contrast the mating of swans is one of the most beautiful rhapsodies of nature. This very afternoon as I walked out of my front door the pair of swans which nest each year above the mill, their lower boundary, were mating. I have seen this on a few occasions and the approach to the actual mating is easily recognized, so it can be photographed or filmed. The birds begin to put their bills into the water in unison, then move alongside one another and gently entwine their graceful necks. The cob then mounts the pen, they mate and, facing one another, both thrust forward, rising in the water as they do so, so that their bills are at full stretch in an ecstatic kiss and their feet only just touch the water. The scene is certainly as beautiful as the ballet but today they uttered only two grunts, unworthy of Tchaikovsky and inferior to the wild musical calls, some say in harmony, of their cousins the Bewicks and Whooper swans, which sound even more melodious and thrilling than the honking of geese.

Swans are supposed to be capable of breaking an arm, but this I doubt. I was meeting a scientist over from Canada at Salisbury station one wet November night. As I drove up the incline there was an even worse than usual traffic jam, so I left my car and walked up the incline to find the ticket collector, whom I knew to be interested in birds, standing beside a swan which had evidently mistaken the wet forecourt for a lake and therefore appeared somewhat bewildered. I told the ticket collector, rather foolishly, that I thought the best procedure would be to catch the bird so that it could be taken to a place of greater safety, as Salisbury station on a wet Friday evening is hazardous to pedestrians and to motorists, as well as to swans. The swan flapped away as I approached, so I ran after it, caught it by the neck and tucked it under my arm in the manner of a pipe major or Alice's flamingoes. The swan had a ring on its leg so my friend the ticket collector made a note of the number, tore the page out of his official book and thrust this into my pocket. Fearing that the swan might damage itself in my car I decided to wait for my friend along with others who were meeting their husbands or wives on the train from London. I felt most self-conscious as I waited and needless to say several of my now ex-friends appeared, to ask foolish questions on the subject of swans for supper and even to make silly references to Leda.

The train was inevitably late and I had to explain to my friend on arrival that I did not usually carry a swan around but thought it safer to release it at home in case it was injured. I squeezed my friend and the swan into the passenger seat with some difficulty, but we arrived home safely and launched the swan on the broadwater below the mill where the road runs alongside the river. During the night there was much flapping as our ill-mannered resident swans drove off this urban intruder. The next morning we watched him walking in a northerly direction across the water meadow above the mill. That was the last I saw of him. Some weeks later I had a letter from the man who had ringed him – in Salisbury!

## DIPPERS AND WADERS

For the last two winters a bittern has appeared in withy beds above the house, although not within view of the mill itself. I feel sure that as nesting birds return each spring to their nesting sites so do winter migrants return to their same feeding grounds. Last year I saw the bittern very close, so wonderfully camouflaged that part of the reed bed itself seemed to rise and flap slowly away. Alas, I doubt whether the boom of the bittern will ever be heard in the valley again as our reed beds are probably too small and, whilst eels are plentiful, grass snakes are now never seen here and frogs have long since vanished. I was given seven frogs and some spawn which I put into a stagnant ditch; whilst scything last summer I narrowly avoided beheading a couple which luckily leapt before I looked. I am trying to get many more to entice a bittern into the garden stream, where once I saw one, when a boy.

I once saw a dipper flying up-river and apparently into the house. I ran downstairs and peered cautiously into the river room but there was no sign. They used to nest on the Palladian bridge at Wilton – there could be no more beautiful site in England, and I believe they still nest on the Nadder and Wylye further up the river. I still hope that with the cascade of water through my mill hatches and the tumbling torrent below I might yet be visited by a pair, always a reminder of spring salmon fishing in Scotland. How they appear to walk along the bottom and why they 'dip' so continuously are pleasantly unresolved mysteries.

Redshank very occasionally appear. They are common below Salisbury but they seem to prefer the Wylye valley to the Upper

Avon. I do not remember that they ever nested here, which is sad as their alarm whistle is splendid to hear. Snipe still drum each spring over the mill – that strange bleat produced by the two extended outer tail feathers as the bird dives towards the ground.

Woodcock sometimes visit the withy bed here and two years ago I filmed one feeding within a few feet of the house, when snow was deep elsewhere. They seem to rest their sensitive bills on the surface to locate the worms before thrusting the bill in, often at an angle rather than vertically, as I had supposed. They are never common here as the ground is unsuitable for them. The presence of woodcock may be seen by observing 'cow pats', since they search these for succulent pink worms leaving unmistakable wedge-shaped marks where they have probed. Several of my friends have witnessed these beautifully camouflaged birds carrying their young; perhaps I have too, for I was driving along a track beside the Spey through a fir plantation when I saw a woodcock rise and fly clumsily away with her legs trailing as if carrying something. She flew only about fifty yards, alighting in the bracken. I then noticed a young woodcock beside the track

and felt sure that the mother would return for it too. Alas, when she returned she settled in the wood behind, and evidently called, as the young bird ran off in her direction. My only doubt as to what I had at first seen came a year or two later when I was staying in Angus. While I was taking my spaniel Drake for a pre-breakfast walk, he flushed a woodcock and three young from the bracken. It was later in the year and the young, which were over half grown and therefore impossible for her to carry, flew only a short distance whilst the hen flew down the ride with legs trailing in exactly the same manner as the Speyside bird – in this case a ploy to entice away both man and dog. The first instance will now remain for me one of many unsolved mysteries of happy memories along with the weight of a large salmon which escaped after an epic battle in Iceland, so nearly won, and the positive identification of a ruff in my garden flushed by a passer-by as I raised my binoculars.

## THE HERON

The heron is a bird beautiful of plumage and graceful in flight but is a voracious and ruthless killer of fish. As a fellow fisherman I am fascinated by heron's varied methods of fishing – sometimes standing motionless waiting for prey to come to them. Their legs are supposed to attract fish but more probably they stir up the mud with their feet in order to bring fish, such as eels questing for food. I have seen them raise their wings high over their heads and dash about on the shallows, usually without success. Whether they believe that shoal fish, like grayling, will seek cover in their shade or whether they try to improve their vision by lessening the reflection I don't know, but I suspect that it is a ploy of novices!

Soon after I came to live here a fine adult heron appeared almost daily with a special fishing technique seemingly derived from watching our local fly fishermen. I named him 'Sir Edward' after the very distinguished Foreign Secretary and dry fly fisherman, Lord Grey of Falloden, author of my favourites *Fly Fishing* and *The Charm of Birds*. I once had the privilege of meeting him –

77

albeit across the river – with my father, who many years before had shared the same syndicate at Itchen Abbas. We met near the place where this winter our bittern has lived for a month, of which he would have greatly approved.

'Sir Edward' would arrive each morning at the bottom of Beat 2 and after surveying the scene, as all good fishermen should, would stalk up the path peering over the reeds into the river. If and when he saw a fish, his body would stiffen and he would approach as close as he dared, remain with beak pointing whilst he adjusted his aim, then plunge in head first and usually emerge rather wet and ruffled with the trout impaled on his beak. I could never see how he got the trout off his beak and into his throat head first, so swift were his movements, but he did not appear to use his feet. He would then saunter on upstream. Once he followed the fishermen's wooded path up a slope until he reached a point immediately opposite my window, which overlooks the mill pool itself. The early morning sun was shining on him and as I watched he spotted a fish some eight feet below. He clambered rather clumsily down, catching his wing on a branch, but this evidently did not frighten the fish. He then paused, aimed and dived from some four feet and with my binoculars I could see that he had speared a fine fat trout of about $1^1/_2$ lbs right through the middle!

I often wonder whether they have some built-in form of radar which locates the fish and calculates allowance for refraction. I saw a heron hasten through some reeds and wade several yards to impale a fish which he could not possibly have seen at the point from which he started. I once saw Sir Edward turn aside from the path, stalk and spear the nest of a mouse – so perhaps sound, or sonar, is also used. On another occasion Sir Edward was stalking as usual upstream when I saw a duck mallard approaching downstream with her brood. To both my surprise and the heron's, the duck saw the crouching bird on the bank above and flew straight at him causing him to stagger back whilst the little convoy passed safely downstream. Herons, when they miss, especially with mice or voles, wash their bills carefully in the river.

The last that was seen of Sir Edward was by a friend who heard loud squawkings and saw a heron being pursued by a peregrine in a northerly direction. Peregrines were not uncommon in those days. I do not think that a single peregrine falcon would kill a heron except perhaps in the air as, once forced down, the powerful beak of the heron would keep the falcon at bay. George Lodge, the distinguished bird artist, naturalist and falconer, once showed me a ring inscribed with the name of a hawking club. He told me that when heron hawking, two peregrines would try to fly above the heron to force it down. The falconer would then run or gallop up, put such a ring on the leg and then release the heron.

# THE RAPTORS

Of the raptors, a pair of hobbies used to nest in the wood a mile upstream but, probably because of the ruthless egg collectors, have not bred here for some years. They do sometimes appear. Their main food is insects, especially moths and cockchafers. Hobbies are mainly crepuscular but I have often watched them catching mayflies in the daytime. They catch them with their feet and then eat them in flight like a chocolate! I once saw a hobby trying to catch a butterfly which escaped by acting like a falling leaf. A party of martins which had taken refuge immediately above the hobby to escape a possible stoop were highly amused by the sight, as hobbies can outfly and catch a swift without too much trouble. Hobbies are most beautiful little falcons and their flight is wonderful to watch.

Swallows can instantly identify a hobby, which is of course essential, and never confuse a kestrel. Our mallard recognized an osprey, which they had never before seen, as harmless to them, whereas a marsh harrier or peregrine would have caused pandemonium. This hereditary recognition is indeed one of the wonders of nature and no doubt a major factor in the survival of certain species.

Buzzards have increased in numbers considerably in recent years and are fairly often to be seen soaring over the nearby 'hanging', as the steep wooded down here is known, especially when the wind striking upwards from the face causes a thermal. Rooks delight to soar there too, tumbling and rising again in their enjoyment of flight.

A pair of kestrels frequent the meadows which abound with mice, shrews, and beetles; a friend who has one which roosts in the creeper on her house tells me these are their principal food. They are wonderful to watch as they hover stationary even in a strong and variable gale, spotting their prey often from a considerable height – a swoop, a short flutter and then supper.

An old elm was the home of tawny owls until the dreaded elm disease struck but they have moved to the old ivy-mantled and pollarded willows beyond the river and sometimes hunt very close to the house. A pair of my favourite barn owls used to float over the meadows like winged ghosts, as the mist rose in the cool evening. Now they are for the time being a memory, but a recorded memory since Robert Bateman, the distinguished and in my view the best wildlife artist in North America, was staying with me, and whilst photographing the mill saw a barn owl alight on the fence just outside. As he stalked it, the bird pounced on a mouse and returned with it to the fence. He painted a picture of this event which hangs in my hall. A single barn owl has been seen not far away so I am hoping that a mate will appear this spring to replace the memory.

I am told that their demise is due in part to eating rats and mice which have been poisoned, and also to their habit of hunting on roads. Their prey is more easily seen, of course, but on the approach of headlights they rise to avoid them and if it should be a tall van they may crash against the unlighted front.

My most distinguished avian visitor appeared when I was looking out of my bedroom window. I saw a large raptor settle on the willow bough which overhung the river from the far side. I always have a pair of binoculars at hand, and spotted the white crown of an adult osprey. Some twenty or more years before one had arrived in the autumn a mile further up the river, and had stayed some weeks, so I know the bird well. I was able to watch the feeding habits of my new arrival throughout the month of October. Each day he settled soon after dawn on the same perch and when he considered that the light was sufficient would launch and circle with laboured flight over the shallows below the island. When his perhaps polarized eyes spotted a fish he would plunge feet first with talons extended and seize the fish, rising with it with some difficulty and, when clear of the water, take it to a dead branch of an oak tree some forty yards back from the river. This, and his roosting perch, were the only branches which he used in a stay of nearly five weeks, by which time his oak dining-room table was so scarred by his talons and rather greedy feeding habits that a white mark was clearly visible from my window. After breakfast

he would depart upstream for a siesta, or perhaps another fishing ground, returning later for supper. Finally he would leave for his roosting branch, which I never discovered.

Once or twice I saw him dive from his perch like a kingfisher, only feet first, but he usually patrolled his beat of about half a mile at a height of about thirty feet, occasionally hovering in clumsy fashion. Often he missed and on one occasion, when the river was high and fast, he failed at first to take off and floated downstream with wings outstretched through a company of amazed mallard before at last, to our relief, taking off successfully. The mallard were curious rather than alarmed and used to gather beneath his perch arguing amongst themselves but, with their in-built recognition charts, as I have said, they identified him as friend and not foe. I even saw him dive through their midst, rising in triumph with a trout and receiving, as it were, a floating ovation – only two rather nervous young ducks jumped in alarm. There was a rumour that our osprey took a trout which was being played by a rival fisherman but I could never substantiate this!

The osprey was an adult and was very tame, so could perhaps have been one from Loch Garten on passage south. How satisfactory that so many pairs now breed in Scotland thanks not only to the efforts of the RSPB but also to many Scottish lairds who protect and welcome them on their estates. A friend who has a small artificial loch in front of his house has to restock it each spring for his ospreys.

When finally he left us I went to look under his oak dining-room table and found there the tails of numerous trout, all of about three quarters of a pound, but no sign of the grayling which abound here. I suppose he shared our preference for trout; also 'ombre', as the French so aptly term grayling, are more difficult to see.

## OTHER BIRDS

I have only one nesting box. A pair of blue tits nest there each year in March, and thereafter there is continuous activity until the end of June as successive tenants fly endlessly back and forth with beakfuls of food for their voracious young.

My bird table collects the usual country diners, especially blue and great tits, which, at times assisted by the elegant nuthatches and that inevitable quarrelsome personality, the robin, eat an amazing quantity of nuts. Coal tits seldom come and longtailed tits do not appreciate the fare which my balcony offers; some winters a marsh or willow tit visit me and even with my 'bird bible', Peterson's *Birds of Europe*, in one hand and the bird only a few feet away, I am quite unable to identify the species; great spotted woodpeckers climb up the posts and peer anxiously over the balcony's edge before visiting the table; green woodpeckers are often in the garden and the little pied woodpecker occasionally appears but never at the table; tree creepers creep up my trees, and the delightful longtailed tits inspect every twig in winter and last spring nested in my bamboos, which, having flowered, all died. The tiny goldcrests seek their food in the ivy-clad alders close to my north windows. Most bird hides are at one level, but my mill allows eye-level observations from tree-top to river.

In early April I always look forward to the arrival of the spring migrants: the little sand martins, the house martins, the swallows and, lastly the screaming swifts, all attracted by the fly life which the river provides. I love to watch the incredible evolutions at great speed of the swifts, with no traffic lanes and yet no collisions, perhaps even more amazing than the flashing white and grey of a flock of waders skimming along the tide's edge, as if controlled by some radar command. So far no swallows or house martins have nested on the mill nor, so far as I remember, did they do so when the mill was operational.

Warblers too should be arriving soon, and the monotonous-voiced cuckoos with their repellent habits, to deposit their eggs in the nests of reed warblers, or whatever species of foster parents may be chosen. I am occasionally plagued by a few starlings but seldom by sparrows, which keep to the villages and farms though these are only a few hundred yards away. Collared doves have the same attraction to towns and villages but a pair usually nests in my garden. Perhaps their call is repetitive but they are beautiful birds and do little or no harm as they seem to prefer seeds to cabbages, unlike the melodious wood pigeons. How strange that these doves should have crossed Europe and continued nor'west to our outer isles in the last twenty years. Sadly, the gentle migrant turtle doves seem to have diminished greatly in numbers since the arrival of their collared cousins. Perhaps because they share the same food, or because of some disease carried by the collared doves, as may be the case with the squirrels. Certainly our beautiful native red squirrels vanishing with the coming of the grey squirrels is a grievous loss.

The sombre little wrens – so modest in colour compared with the blue fairy wrens of Australia – are for ever searching the creepers on the balcony for insects and find their way into the conservatory to help rid me of aphids and other pests. In a hard winter they will roost communally for warmth and I once counted over thirty seeking shelter in the corner of my roof – there must have been many more since I could only see those approaching from the south. I fear that their entrance was eventually blocked one spring by a pair of nesting great tits. They search too for food amongst the roots and weeds just above the river level.

Of the spring birds, I welcome especially the common flycatchers and love to watch them as they flit back and forth between posts like tennis balls, catching their insects en route. Their nests are so obvious that even I can locate them. A hen blackbird with a white spot on her back stayed with me for four years until last winter and was usually to be seen on the north side of the house, occupying as far as I could see only a small territory. Dunnocks or hedge sparrows feed in their peaceful and quiet way on the balcony in company with the stumpy greenfinches, whose golden cousins like to feed on the thistledown in autumn, while the bullfinches remove buds in wanton and purposeless fashion from the fruit trees.

Stone curlews which were so common before the war on our downlands and stoney arable fields are now seldom heard or seen; however they still nest at Porton Down near our bustard pen. This ancient stretch of downland has neither been sprayed nor ploughed and thus affords sanctuary to many species of fauna and flora which the authorities at Porton carefully preserve. Great bustards are recorded in *Birds of Wiltshire* as having nested in the water meadows not far from the mill about 1810 and it is our endeavour to restore these great birds – possibly the heaviest of all flying land birds – to Salisbury Plain. A bustard still proudly surmounts the Wiltshire coat of arms.

Quail, which like the same habitat as bustards and stone curlew, used to nest nearby most years and now show some signs of returning; they were heard calling near our bustard pen the last two springs and last autumn I saw another a few miles away. Sadly, as I mentioned earlier, the landrail or more aptly named corncrake has long since vanished from the harvest scene.

Whilst this chronicle mourns the passing of some species, many others have increased in numbers and I would judge that there are now as many resident and migrant birds in and around our beautiful valley as before the war. By late April of this year, 1984, very few of the swallow family had arrived due, I feared, to the report I had read of a very serious drought covering a large area of North Africa which cut off insect food supplies for many migrants on their long journey.

Swifts, no doubt, would overfly the danger area and so arrive in their usual numbers but I had seen only one swallow and no martins. I gazed south each day, looking down the river and hoping to see swallows and martins hunting insects above the river as in previous years, but with increasing fears for their fate.

Now, in late spring the swallow family have eventually arrived. Swifts are as numerous as ever, swallows and house martins far fewer than usual and the graceful and delicate little sand martins scarce indeed. Perhaps a fine late spring and summer will allow many second nests and thus restore the numbers of this beautiful family.

# The Mammals

Of the mammals which live beside the river and streams, the water voles are delightful – like Ratty of *The Wind in the Willows*. These fat short-tailed and short-sighted little rodents hibernate for part of the winter but on milder days emerge for short, sleepy excursions on the river, perhaps to visit their friends, whilst in spring a 'ferry service' crosses and re-crosses every few minutes below the daffodil promontory, operated by Charon, the ferryman.

In the summer water voles are always to be seen, either crossing the stream or sitting comfortably on the weeds, munching thoughtfully at a piece of weed which they clutch in their little hands. The rack above the mill is an endless source of fresh weed, and there one sits for breakfast, lunch or tea. They are entirely vegetarian, and sometimes eat grass on the lawn. The only damage they do is to the river banks where they tunnel, but who would resent this from such delightful neighbours? In autumn they drag weed into their burrows for winter feed. I once watched one swim out to a passing weed raft, seize a succulent strand and start off to drag it ashore. When the vole found that the raft was dragging it downstream it swam back and bit off the strand from the raft, returning with it in triumph to the bank.

Recently Rodger McPhail was staying with me and as we sat on the balcony, a kestrel swooped down at a young water vole. Fortunately she missed. Although water voles are very short-sighted and seem to be slow, I saw a young vole leap clean out of the water when playing with his brother – so in emergencies they can take swift evasive action. The games that many young animals play are surely training for such dangers. Kestrels usually concentrate on mice and beetles but, oh, little and grown-up water voles do please beware of the grey heron.

I have been endeavouring to study more closely the habits, numbers and homes of these delightful creatures. Below the house on the left bank are three holes with a fourth at Daffodil Promontory. No. 1 is inhabited by a one-year-old; at No. 2 resides Grandma Moses. I feared that both had been devoured by the grey heron who invaded the stream during the hard weather, and whilst I never actually witnessed him killing a vole I have absolutely no doubt that he does so; I have watched him with bill pointing directly at a hole, and with his length of legs and neck he can strike a fish or vole with rapier thrust at a distance of at least a yard. During a mild spell in early February, however, I was delighted to see Grandma Moses' grizzled muzzle peer nervously from her home, after her long winter's siesta. No. 3 seems at present to be empty whilst at No. 4 resides Charon. By the little side stream there are at least three more homes, whilst just above the weed rack lives a very fat vole who even in winter spends much time on the weed rack – he believes it was put there for the special purpose of catching the weed and sometimes watercress which float down river for his insatiable appetite.

Another water vole lives under the roots of a fallen willow and yet another above the point where the mill stream rejoins the main river.

Whilst the voles often consort together on the river and certainly at times inhabit the same hole in winter they do appear to prefer their own company. All their homes have backdoor emergency exits into the river itself, usually below the surface. The young do not seem to be allowed out until they are about a quarter grown and I have never seen more than two playing together, although voles are thought to have families of four or even five.

Birds and animals which are tame are especially delightful. Water voles apparently both deaf and blind to our presence, can be approached to within a few feet and their little noses, twitching as they eat, seem unable to detect our strong human scent.

Stoats and weasels hunt in my garden at times but although they follow the river bank I have never seen one kill a water vole, nor seen a vole which has been so killed. Certainly they are quick to dive at the last second if danger threatens – or perhaps the stoat family prefer rabbits! Let us hope that their mink cousins do not invade our river here. On the stretch of the Wylye colonized by those alien and ruthless killers, the delightful water voles have all been killed, yet so-called Animal Liberators released several hundred mink from a farm there and attempted to do so on the Upper Avon. How can these so-called friends of animals justify the ensuing slaughter of ducklings and voles, and destruction of small birds' eggs? A farmer on the Frome went to feed his twenty chickens only to find that all except one had been killed and a blood-stained and sated mink slept amidst their corpses.

How different from the voles are the long-tailed nocturnal prowlers, the evil grey rats. These robbers of eggs, killers of small birds and carriers of disease, often follow the banks of rivers and I am always amazed by the number of people who do not know the difference between these miscreants and water voles.

I once found a number of rats robbing my bird table. I bought a dozen packets of rat poison and was assured that each packet taken meant a dead rat. I was careful to lay the packets face down so that these super rats could not read the poison notice, and I put them out of reach of my dog. After ten packets had been taken, and ten rats supposedly killed, I noticed a trail of the powder leading to, and under, the tool shed door. There in the corner I found all ten packets made into a nest by the super rat or ratess!

Otters, our delightful fellow fishers, were always present after the war. Pike seemed to be frequent prey although eels are their favourite food – smoked eel is certainly one of mine. Whilst we still have a pair of otters on the Frome where I fish, I have seen no signs of any here myself. However the Piscatorial Society's excellent fishing keeper reported that one had been seen by a member and that he himself had heard one last season. So perhaps they may return and I may yet see that unmistakable flat head coming up the stream towards me, as the light fades. Last season, whilst fishing the evening rise on the Wylye, I saw a long humped, serpentine creature swimming fast upstream in the dark waters under the far bank. I have read of otter cubs holding onto the rudder (tail) of their mother or the cub in front and believe that I was fortunate enough to have witnessed this rare event. The fishing-keeper, a lover of otters like most sensible keepers, confirmed later that there was a bitch otter with four cubs – a happy event.

> *Where the night air cools on the trout-ringed pools*
> *Where the otter whistles his mate.*
>
> (Rudyard Kipling)

Roe, those most beautiful and graceful members of the deer family, especially in their red summer coats, were never seen before the war. Now they are common. I have seen one cross the river below the island, and I saw the tracks of another just above the mill a week or so ago. They seem to like the seclusion of withy beds and they also unfortunately, like roses. I read that a sack soaked in creosote will keep them away as their acute nostrils can no longer then detect human scent. A friend hung human hair on the fenceposts. I am happy to report that this came from her hairdresser and were not the scalps of her enemies. The result was most successful and certainly less expensive than the nine foot wall mentioned in one book of gardening lore as the only real cure.

On summer evenings bats congregate around the mill feeding on the many flies which have survived the depredation of the daylight feeders, the swifts, swallows, martins and so many

others, and to a small extent the trout themselves. Peter Scott told me that these are Daubenton's, or water bats, which would certainly make sense!

Of the smaller mammals, I caught in my conservatory a most beautiful mouse which had been doing considerable damage. It had large, bulging eyes and soft velvety ears, a very long tail and pale yellow neck. This proved to be the nocturnal and very agile yellow-necked field mouse. I did not try to catch another. As to whether water shrews are present in the streams, I have never been sure. At times I have seen little 'plops' as I walk along the banks but have never been quick enough to see the little 'plopper', which seems to vanish under the water. Wild untamed common shrews, field mice and many other little creatures abound, the prey of the 'windhover' by day and the tawny owls by night, and perhaps soon once again supper for the beautiful feathered barn owls. I was spraying an ailing magnolia on one occasion when a shrew emerged scratching itself like a dog with fleas – she then vanished beneath the roots, to emerge carrying a baby, like a spaniel retrieving a rabbit, and carried it to a neighbouring and healthy shrub. She returned to collect all her family from almost beneath my feet.

What a pity that the moles should throw up so many little mountains on the lawns and flowerbeds in their unending quest for worms; but I am sad when I see the trap has sprung. I met one crossing the road towards my garden and turned him round so that he faced again the garden from whence he had come – most un-neighbourly behaviour.

Those curious and attractive hedgehogs visit me at times but, being nocturnal, are seldom seen. My nephew who lives high above the village saw one walking about with what appeared to be a white top hat. He went to investigate and Mrs Tiggywinkle was able to explain that she was eating yogurt from a carton when her head had become stuck so she wandered around hoping to find a helpful friend. She was given bread and milk for which she returned each night and as far as anyone knows lived happily, if not ever after, at least for as long as hedgehogs live.

A few days ago a friend was driving away when he stopped the car. I went to investigate and he pointed to a young hedgehog evidently out for an evening stroll advancing down the drive, past the car, through my legs and on towards the river. I ran to get a landing net and, being unsure as to his swimming capabilities, caught him as he was about to enter the river. I carried him back to the 'butterfly garden', from whence he seemed to have come. As soon as I put him down he set off once more on his evening excursion, down the drive then right along the river path beneath the weeping willows, whilst I walked anxiously beside him. I was expecting that we would arrive at his mother's daytime retreat, especially when after fifty yards he suddenly turned into the rushes beside the river. However he continued on his strange journey as if in a trance, launching himself into the water where, in this new element, his rotund body floated whilst his little legs continued their terrestrial trot. He managed to attain a satisfactory propulsion but no sense of direction, lacking the streamlined otter's webbed hind feet and rudder. Thus poor Master Tiggy-winkle – or perhaps Miss – was at the mercy of the stream, and I ran ahead like a distraught duck whose brood is in danger. I reached the hatches in time to climb out along the weed rack and net the now bedraggled little animal before he was sucked into the mill pool.

I put him into a box and gave him an early supper of bread and milk which he ignored. The light was fading, and knowing how often tame animals meet with disaster I returned him to the butterfly garden and covered him with hay. The next day he was gone.

# Epilogue

These creatures and plants which are my neighbours will only survive for future generations if the river runs full, clear and unpolluted. The number and variety of birds and fish are there because of the abundance of fly life on which their young depend.

I apologize for my frequent diversions, often out of context. As I write I continue to see things of interest, so let us conclude with my latest and nearest observation.

This week I was wandering rather aimlessly just above the mill when I espied a strange, rather bedraggled something swimming with great effort and little progress towards the bank. I sat down to watch and as it approached I saw that it was Mole, his little forepaws or, more aptly, fore-claws flailing the water for which they were so ill adapted whilst his hind legs hung down like those of a child learning to swim. Eventually he reached the steep and overhung bank just below me. I was just pondering how he would surmount this obstacle when, after a surprisingly short time, I noticed the grass opening beside me and Mole surfaced through the chalk, the questing nose and flailing claws back in their own element. I half expected him to say 'Have you seen Ratty?' – as indeed I had, nibbling weeds held between his paws a few yards downstream – when Mole vanished underground leaving as little trace as Ratty when he dives.

Kingfisher Mill may not be as grand as Toad Hall but at such times it has the same dream-like quality, especially when the white mists enshroud the valley on a still summer's eve and then 'leave the world to darkness and to me'.

OSPREY PERCH

RIVER

KINGFISHER PERCH & NESTING BANK

DAFFODIL PROMONTORY

LITTLE AMAZON

BUTTERFLY G...

BEEHIVES

ISABELLA'S ORCHARD

# KINGFISHER MILL